The Return to the Sinister Universe:

The Immaculate Collection

Greg Feild

September 5, 2018

Abstract:

This book is a compilation of my final four efforts on The Universal Model of Our Sinister Universe.

A quantum mechanical theory of everything

On interaction

On rotation

Revenge of the sinister universe

A Quantum Mechanical Theory of Everything

Greg Feild

November 5, 2017

About the author:

Greg Feild is a physicist.
His mission is to put the *physical* back into physics.

His motivation is to understand the real world.

I suppose it is tempting, if the only tool you have is a hammer, to treat everything as if it were a nail.

-- Abraham Maslow
The Psychology of Science

Abstract:

In this book, we offer a final account of
The Universal Model of Our Sinister Universe.

The main purpose is to summarize the theory
without the distraction of the numerous errors
committed along the way.

This book is not an exegesis or exposition,
but a synthesis of our several speculations
spanning thirteen recent papers, for a final,
wart free, theory of everything!

Submitted for your approval ...

Sapere aude! -- Horace

Time is the formal a priori condition of all phenomena whatsoever. Space, as the pure form of external intuition, is limited as a condition a priori to external phenomena alone.

If I can say a priori, "all outward phenomena are in space, and determined a priori according to the relations of space", I can also, from the principle of the internal sense, affirm universally, "all phenomena in general, that is, all objects of the senses, are in time, and stand necessarily in relations of time."

-- Immanuel Kant
Critique of Pure Reason

Preface:

This book contains (only a) few new ideas and not a lot of new text.

Most of the text has be been recycled from previous books.

The purpose of this book is to provide a summary of our new model, separating the wheat from the chaff.

Please enjoy!

Greg F.

Feeling Gravity's Pull:

Reason had harnessed the tame
Holding the sky in their arms

Gravity pulls me down

-- R.E.M

The universal model:

Until our new collection of theories, people did not understand the origin of particle mass, why mass and energy are equivalent, how accelerated particles acquire mass-energy, the nature of antimatter, the nature of particle-antiparticle annihilation, particle spin, the particle family hierarchy, muon decay, gravity, nor last and what *should be the very least*; the fundamental physical mechanisms underlying particle interaction.

Yet theorists speculate wildly. Experimentalists look for "new physics" !

Today's theoretical approach is all about mucking around with mathematical models; without an underlying physical model, it seems. If you get the math right, the correct model is sure to follow! But whence this 'math'? Monkeys at typewriters spring to mind ...

Physics is mathematical because we measure things and then try to organize and understand them in a quantitative way.

First you do the physics, then you do the math. Lather. Rinse. <u>Repeat</u>.

If you can *later* formalize your math into something beautiful. Great!

Lately, physicists have been putting the cart before the horse.

Actually, the cart has been abandoned in a ditch!

I *only* scold because certain physicists have been *shamelessly* airing our dirty laundry, *and ignorance*, turning a tidy profit, all the while; probably in the name of 'public education and outreach'. Unfortunately, they speak *complete nonsense*, undermining public confidence in science in general, and providing needless fodder for the fanatic followers of the latest fads and fashions.

Who is in charge of these people ?

I'm not angry anymore!   :)

Just disappointed.

I'm off to see my psychic astrologer to have my quantum hologram unentangled.

Introduction:

Scientific theories must be rational and they must be logical. Scientific theories must conform to human reason and to common sense.

Reason is all we have to discern truth from falsehood; reality from fantasy; fact from fiction.

Curved spacetime, extra dimensions, extra universes, parallel worlds, collapsing wave functions, non-locality; none of these ideas can be reconciled with human reason or sense.

For this reason alone, they must be dismissed out of hand and with extreme prejudice.

These ideas are silly.

It is the 21st century. There is no room for magical thinking, supernatural entities, or superstition in science.

Space is three dimensional. Elementary particles are solid units of matter. All interaction is mechanical and deterministic. There is only one universe.   Time flows forward.

When did humankind lose its way?

When did the physicist become ontologist?
Who allowed such a transgression?

Where were the *philosophers*?

Will human beings ever mature?
Will human nature ever change?

Or, will there *always* be beasties?

Oy vey.

The special theory of relativity:

The erroneous argument and conclusion of the special theory of relativity may be stated as:

i) Since transformations between inertial reference frames are no longer Galilean,

ii) Space and time must no longer be Newtonian.

Expressed in this manner, the argument looks weak, if not fallacious, already.

The inertial reference frames between which we must make non-Galilean transformations are themselves Newtonian reference frames, each and every one. Time and space appear the same in all inertial reference frames; flat, isotropic, homogeneous, etc. Despite its name, the special theory of relativity implicitly prefers the observer at rest and makes a special case of the observer in motion. Of course, the whole point is the observer "does not know" he is motion (without reference to something else), and time and space certainly "don't know" whether an observer is in motion. Only *relative motion between interacting particles* matter.

Choice of reference frame is a matter of convenience in bookkeeping. Space is space.

(Why figure in base twelve when you have ten fingers and toes?)

The correct argument and conclusion from the special theory of relativity is as follows:

i) Since the mass and energy of a particle are equivalent, and the energy of a particle is a *nonlinear* function of particle velocity,

ii) We may no longer make Galilean transformations between inertial reference frames.

Time and space "don't know from" your coordinate transformations!

Oy vey ist mir.

The universal principle of equivalence:

In our first book "A Quantum Mechanical Theory of Gravitational Interactions",
we began our enquiries by proposing a modified principle of equivalence;

inertial mass = gravitational mass = gravitational charge   (1)

Our recent investigations suggest an even more general expression for this principle,
which we shall call the 'universal principle of equivalence'.

relativistic inertial mass = relativistic gravitational mass = gravitational charge

== *the fundamental universal coupling charge*   (2)

In our new model, the "relativistic" particle mass is the gravitational charge,
the "strong force" charge, the weak charge, and ultimately, the electric charge as well!

The universal theory of relativity:

Why do we use "scare quotes" when we say the "relativistic" mass?

The postulates of the special theory of relativity are;

i) Physical laws are the same in all inertial reference frames.
ii) The speed of light is a universal constant in any reference frame.

As an alternative, we propose the postulates of 'the *universal* theory of relativity';

i) Particle interactions occur at the speed of light in any reference frame.
ii) Particle interactions obey the $1/R^2$ law in any reference frame.

iii) $F = d\mathbf{p}/dt$

In an inertial reference frame, I believe, these three postulates should lead one to the
familiar expressions for the relativistic mass and momentum, *without* reference to,
or resorting to, coordinate transformations.

The familiar Lorentz transformations would then follow naturally as a consequence.

Inertial reference frames:

In our model, the total relativistic mass of an interacting object is a function of *all* relative velocities between the object and a second interacting object.

The total 'relativistic' velocity between two bodies now includes contributions from angular velocity, and 'intrinsic', or absolute, *rotational velocity*, in addition to the usual rectilinear velocity.

Because spin is an inherent component of our theory, and because everything is spinning, there can be no inertial reference frames, even in principle!

We suggest the inertial observer (who is always 'at rest') reference their inertial coordinate system to the " 'fixed background' of 'empty space' ". (You may rearrange the "scare quotes" as you'd like!)

The 'fixed stars' are no longer fixed, nor must we worry about their influencing our measurements. The stars will either be part of our study, or too far away to matter.

The universal reference frame:

In our model, (the background of) space is fixed and immutable; 'flat', homogeneous, isotropic; Euclidean. Empty. Void.

An inertial observer is *defined* to be absolutely at rest, against the fixed background of space, if they measure the accepted, absolute value for a particular and well known wavelength of the cosmic background radiation.

An observer in a *moving* inertial reference frame, could then determine their absolute velocity relative to the fixed background of space, by measuring the Doppler shift of the cosmic microwave background.

In our model, the cosmic microwave background consists of gravitational waves (photons). Residue from every mass ever accelerated, anywhere (within "range" of our 'sector'), ever!

The speed of light:

    The speed of light is the speed of particle interactions. The speed of interaction between particles is independent of the velocity of the particles, and if we feel compelled to introduce an observer, the choice of reference frame. This attests to the fixity of space.

    However, the *frequency* (and thus 'strength') of particle interactions is dependent on the <u>relative velocity</u> of the particles. This *is* physics.

    Time may be not be absolute, but in any given inertial reference frame, time is fixed and invariable, ticking away regularly, ceaselessly, and eternally. (i.e. Newtonian.)

    This is all that matters, and all we can ask for!

    Why can't a particle travel faster than the speed of light?

    A particle cannot exceed the speed of light because it cannot exceed the speed of the force (or the source) causing it to accelerate!

    Classically, we can imagine placing an electron in a constant electric field of infinite extent. This field would exert a constant force on the electron and, in principle, we could accelerate the electron to any speed we'd like.

    However, fields are not real.

    Particle acceleration is due to particle interactions. Our electron's acceleration is actually caused by the exchange of virtual photons with electrons on the surface of a capacitor plate.

    The faster and farther our electron is accelerated, the farther the next virtual photons have 'to travel' to give our electron its next boost. The $1/R^2$ law in action!

    We can employ similar arguments to explain why all observers measure the same value for the speed of light. All particle interactions occur at the speed of light in any reference frame.

    Only the fact that the different observers measure different frequencies for said light, is keeping our minds from totally exploding right now!

Fields:

In our new universal model of the world, there is one fundamental force responsible for all particle interactions.

We continue to call this one, single, elementary, force electromagnetism, and the corresponding field the electromagnetic field.

Our new electromagnetic field interacts with the 'total coupling charge' of a particle; the sum of a 'mass dependent' electric charge and the relativistic mass of the particle, with the appropriate proportionality factors (e.g. alpha, G, etc.) applied to each term.

In our theory, the electromagnetic field is solely a mathematical field describing the interaction between particles, and *does not* carry energy or momentum.

That is the job of the photon; be it virtual or real.

It does not make sense (or, at least, it is not operationally useful) to talk about the force field or potential field of a single particle; its strength, how far it extends in space, etc.; without the presence of at least one other particle somewhere in the universe!

In the field model, a two particle universe would have an infinite amount of energy all stored in these magical fields, and all arising from two little electrons.

We conclude that a particle need not, and does not, create a force field somewhere where some other particle is not, and in fact may never be.

People get so caught up in their mathematical models that they totally *forget* the basic underlying, original, and guiding thoughts and principles that inspired them.

The first rule, and overarching principle of the universal model is - you cannot do physics with just one particle!

For every action there is an equal and opposite reaction.

Classical field theory essentially throws out Newton's third law, especially when one object is considered a fixed center of force for one or several other objects. The fixed center does not move, or 'recoil'. In addition, all other objects are considered as 'passive' participants in the interaction. Finally, the time delay between any change in the source (e.g. if it were allowed to move) and the subsequent motion of the objects subject to the force, renders the notion of action and reaction logically impossible. Most people are (at least implicitly) aware of these assumptions, but they do not realize just how much damage they do!

In our model, objects are constantly exchanging energy and momentum via virtual photon exchange. Hence, all interacting objects move synchronously, and together, and all at the same time!

In this picture, Newtonian gravity *can be* considered as instantaneous, since celestial bodies conspire to move synchronously due to their *mutual* gravitational interaction.

In our theory, the motion of mass is the source of all interaction. This means *all* relative motion. Particles cannot tell the difference!

When you designate, and then constrain, a force center, be it a proton or the sun, you lose a small, but important contribution to the interaction and the overall motion of the system.

Fields are not real. They are mathematical models with limitations, and they are not real physical entities existing in time and space.

No fields means; no retarded potentials, no waves traveling backwards in time, and no infinite energy sums

Fields: good riddance!

The $1/R^2$ law:

All matter interactions are due to real photon emission or virtual photon exchange.

Both of these manifestations of the electromagnetic interaction follow the $1/R^2$ law of diminishing returns.

The mechanism for the $1/R^2$ behavior is completely different for the two cases, although the reason for the behavior is exactly the same! Both are dependent on space being flat and three dimensional.

For real photons, the intensity falls off as $1/R^2$ because the net flux of real particles (or energy and momentum) through a sphere of any radius surrounding the source, must be conserved.

For virtual photons, the strength of an interaction between any two particles diminishes as $1/R^2$ as a consequence of the uncertainty principle.

In either case, space must be Euclidean.

Force and acceleration:

The underlying assumption of the general theory of relativity is that acceleration should not be considered a special state of motion.

The problem with this idea is that without acceleration, there is really be nothing to talk about!

Explaining acceleration (or more generally, rates of change of states) is the *raison d'etre* of physics.

Furthermore, acceleration is not relative like velocity. An accelerating object will have some acceleration in any inertial (i.e. uniformly moving) reference frame. Observers in different reference frames will compute different forces, velocities, etc., but energy and momentum will be conserved in all inertial frames.

Acceleration is *absolute motion* and is what physics must explain.

In our universal model, angular motion, such as planetary spin, is also considered acceleration and should be regarded as absolute motion. Consider two spinning bodies orbiting one another. One cannot find a reference frame where someone isn't spinning!

A model where everyone is at rest, everywhere, with no established, or prefered reference frame(s), not only removes time from the universe, but also renders the very idea of motion impossible to talk about in any meaningful way.

And yet, things move . . .

It is time to bring back motion and rates of change!

Time and space:

Time and space are conceptions of the human understanding, *and not* perceptions of the senses, representing externally existing physical realities.

There is no time or space beyond our propensity to note the rates of change of states, both conscious "states"; time, and motion; the changes in the relative positions of a system of interacting objects.

This was demonstrated by the much disparaged and maligned *thinker*, Immanuel Kant, over three hundred years ago!

Time and space are emergent properties (to use the popular parlance) of the relative motion between objects. Time and space arise from our need to assign positions and velocities to objects in our study of dynamics.

No objects; no space. No motion; no need for time!
Know objects; know space. (It had to be done!)

In any given inertial reference frame, it is appropriate to define time and space as Newtonian. We have a fixed three dimensional space and a ticking clock.

The only new feature we impose is that a particle's mass is now dependent on its velocity. However, this is *not* a comment on the nature of time and space.

Particles 'lose and gain' 'kinetic energy' during an interaction and this is reflected in the particle's varying mass-energy.

We work with the Lorentz transformations and write our four vectors in Minkowski space because a particle's mass-energy is a nonlinear function of the the velocity of the particle, *even in* our ordinary, everyday, reference frames of three dimensional Newtonian space and time. (This approach is in accord with Kant's metaphysical model of space and time.)

Now, the general theory of relativity is not just a "theory", but a *metaphysical* model of the world; a model within which people 'do physics' as conceived and prescribed by the concepts of the model.

Unfortunately, considered as either a physical theory or a metaphysical model, the general theory of relativity is incorrect. (It's OK!)

It is incorrect about the nature of time, space, motion, and acceleration; as well as the fundamental, irreducible roles of relative and absolute motion; *and* acceleration in particle interactions.

As a metaphysical model, it is not only incorrect, but *un*-correct, as it has led researchers (and not only in the field of physics) down the wrong path.

As it turns out, time and space are 'just normal'. Time and space are just as you and I experience them and inferentially define them every day.

Time and space are 'just normal'; just as Isaac Newton envisioned.

Space just sits there and time ticks.

The choice of a particular inertial reference frame is *irrelevant* for the proper description of the interaction between particles, and this choice does not affect, or have an influence on, the intrinsic nature of time or space, or on the "behavior" of time and space.

Over the last century, the awe filled allegiance and fanatical fidelity to the general theory of relativity, has been a constant burr in the ass of progress.

The theory 'adequately' explains the bending of starlight by the sun and the perihelion of Mercury. On the other hand, the theory predicts curved and stretchy spacetime, the big bang, wormholes, singularities, dark matter, dark energy, acceleration (twice!), and many other *untenable ideas*.

Let it go. It is time.   It's alright.   :)

The Lorentz force:

It is the premise of our model that the motion of mass is the basis of all interaction and all physical phenomena. This assumption was necessary in order to afford the newly massive neutrino a magnetic moment.

For this reason, we find the charge to mass ratio, e/m_rest, of a *charged* particle to be the proper and fundamental coupling *constant* of electromagnetism (long, a popular idea, amongst many!), rather than the electric charge, e, alone, and the *relativistic* mass of the particle to be the actual coupling *charge*.

So, in our model, the complete relativistic Lorentz force on a particle due to a specified distribution of charge and mass is given by

$$\mathbf{F} = (me')*\mathbf{E} + (me')*\mathbf{v}\times\mathbf{B} - m*\mathbf{F\_g} - m*(\mathbf{v}\times\mathbf{B\_g}) \qquad (3)$$

where m is the relativistic mass of the particle, e' is the charge to mass ratio of the particle

$$e' = e/m\_rest \qquad (4)$$

and m_rest is the rest mass of the particle.

The term F_g is the Newtonian force due to gravity and is calculated in the usual way (as tiny as it may be!)

$$F\_g = G*m1/R^2 \qquad (5)$$

where R is the distance between our particle and a gravitational source charge of relativistic mass m1.

The term B_g (very tiny!) is the 'gravitational magnetic field' vector, analogous to the electromagnetic term B, and is calculated in a similar way;

B_g = (G/c^2)*m1*(**v1xR**)/R^3   (6)

where v1 is the velocity of the source particle. The constant giving the strength of the force, G/c^2, is chosen to give our resulting gravitational waves the speed of light.

Equation (3) for the total force on a particle is manifestly Lorentz invariant due to the common mass factor appearing in all four terms. The electromagnetic terms reduce to the usual ('classical') expression in the low particle velocity limit, where m = m_rest.

The original motivation for 'symmetrizing' the classical Lorentz force with respect to gravity was to prop up a quite separate hypothesis, which was that the neutrino had a magnetic moment, even though the magnetic moment arises solely from the 'rotating' motion of its mass.

However, this symmetrization allows for a classical description of gravitational waves *and* bolsters our conclusion from quite separate arguments (1) concerning a gravitationally bound state of two neutrinos, that the graviton is a massless, spin 1, boson just like the photon.

In fact, we were forced to conclude that the graviton and the photon were the same particle for lack of any distinguishing characteristics! (3).

The gravitational waves would continue the electromagnetic spectrum just where the radio waves start to fade away ...

As we constructed our model, we found (or required, demanded) that the gravitational and electromagnetic interactions have the same structure, mechanism, and behavior; the only difference being the value of the electric charge.

In addition, we supposed (postulated) that the gravitational and electromagnetic interactions have the same structure, mechanism, and behavior for both macroscopic and microscopic systems; *the only difference being quantization.*

In this spirit let's have a closer look at our new Lorentz force! Let's rewrite equation (3) as

F = m*(e/m_rest)***E** + m*(e/m_rest)*(**vxB**) + m***F_g** + m*(**vxB_g**)   (7)

where we are not going to worry about the relative plus and minus signs between the electromagnetic and gravitational terms for the moment.

Factoring out the constant coupling charge, e/m_rest, we have

$$F = (e/m\_rest)(m*E + m*(v \times B)) + (m*F\_g + m*(v \times B\_g)) \qquad (8)$$

We see that the electromagnetic and gravitational forces now have exactly the same form and they both depend on and vary with the total relativistic mass energy of a particle.

For the electromagnetic terms, e/m_rest is *the* fundamental coupling *constant*.

The charge to rest mass ratio of a particle is now the fixed fundamental coupling constant, rather than e or alpha, and needn't be involved in calculations except as a multiplicative factor.

We note, when m = m_rest, the electromagnetic and gravitational forces are totally 'decoupled'.

Now, let's take a closer look at the new mass dependence of the Lorentz force as described by equation (8). For simplicity, we will focus only on the two electromagnetic terms. The force on a massive charged particle (e.g. an electron) due to a specified distribution of mass and charge is then

$$F = (e/m_e)(m*E + m*(v \times B)) \qquad (9)$$

If we consider the interaction between two identical electrons, we must assume one provides the field for the other, and equation (9) becomes

$$F = (e/m_e)(m_1*E + m_1*(v_1 \times B)) \qquad (10)$$

Using the well known equations for **E** and **B** due to the second electron, we have

$$F = (e/m_e)^2 (1/R^2)(m_1 m_2 (1/4\pi\varepsilon)r + p_1 \times p_2 \times r (\mu/4\pi)) \qquad (11)$$

Some factoring yields

$$F = (e/m_e)^2 (c/R)^2 (\mu/4\pi)(m_1 m_2 r + (1/c^2)(p_1 \times p_2 \times r)) \qquad (12)$$

For non-relativistic interactions, we can solve the relationship

$$E = p^2/2m \qquad (13)$$

for the mass of our two electrons, finally yielding

$$F_{1,2} = (e/m_e)^2 (c/R)^2 (\mu/4\pi)(p_1^2 p_2^2/4E_1 E_2 + (1/c^2)p_1 p_2 \sin\theta) \tag{14}$$

assuming our two electrons are moving parallel to one another.

The purpose of this exercise was to demonstrate that the classical Lorentz force between two particles can be formulated solely in terms of their energy and momentum.

The classical coupling strength is now given by our new universal coupling parameter, $(e/m\_e)^2$, which can be carried over unscathed and unmolested into quantum mechanical calculations!

We now reintroduce the gravitational terms into equation (12) and find the the complete, 'classical', relativistic, Lorentz force between two identical electrons

$$\mathbf{F} = (G/c^2 - (e/m_e)^2(\mu/4\pi))(c/R)^2)(m_1 m_2 \mathbf{r} + (1/c^2)(\mathbf{p_1 \times p_2 \times r})) \tag{15}$$

Equation (15) demonstrates the *equivalence* of inertial and gravitational mass.

We note that the factor of $(1/c)^2$ dampens the gravitational interaction relative to the electromagnetic interaction. In addition, both the gravitational and electromagnetic 'magnetic' terms are dampened by a factor of $(1/c)^2$ relative to their respective static forces.

Also, notice that the factor of epsilon_0 is 'gone', and we are left with two coupling constants, G and e/m_rest, and a 'space factor', $\mu/4\pi$. The evolution of the force is totally described by the energy and momentum of the two electrons. The factor $(c/R)^2$ describes the time dependence of the interaction, and the time t is *common* to both electrons.

The particles act on one another. There is no 'source' charge.
This description contains no electric or magnetic fields!
All the energy and momentum of the system is carried by the particles.

Nowadays, we characterize this interaction as two electrons trading energy and momentum by virtual photon exchange. Exchange is the key word. There is no emitter and no receiver; just a continual flux of virtual photons between the two.

Actually, a picture I like better, is of one of a continuous virtual photon, constantly changing 'frequency' or 'mass' as the interaction evolves.

Let's look again at the force between two identical electrons as described by equation (15).

We note, of course, that $\mathbf{F_1} = -\mathbf{F_2}$. Newton's second law tells us

$$\mathbf{F_1} = m_1 \mathbf{a_1} \qquad (16)$$

If we compare equations (15) and (16) we can see that the acceleration of a particle is *independent of its mass*.

$$\mathbf{a_1} = \text{Function}(m_2, \mathbf{R}) \qquad (17)$$

This is a general result that we used to assume applied only to the gravitational interaction (and is used as an argument in favor of general relativity).

In addition, the relative strengths of the four terms in equation (15), or the 'four forces of classical physics', are approximately as follows;

electricity = 1 ; magnetism ~ $1/c^2$; gravity ~ G/c^2 ; magnetic gravity ~ G/c^4

I think Maxwell would approve!; except that we have no more need for his fields.

The two electrons exert equal and opposite forces on one another during the interaction, and the evolution of the force is completely described by the (variable) mass-energy of the two electrons, $m_1(\mathbf{r_1}(t))$, $m_2(\mathbf{r_2}(t))$, where the time, t, is *common* to both electrons.

Of course, without the fields there is no mathematical or physical mechanism to explain the interaction of these two particles 'at a distance'.

We like the idea of 'one virtual photon' (which, of course, is a discovery of the field theory model) constantly coupling the particles; a time varying conduit for energy and momentum exchange. Can we infer or derive the virtual photon without resort to field theory?

We will defer this investigation for now.

Let's look at our 'new' Lorentz force in a little more detail. If we define

$$K == (G/c^2 - (e/m_e)^2(\mu/4\pi)) \qquad (18)$$

then we can write equation (15) as

$$\mathbf{F} = K*(c/R)^2(m_1 m_2 \mathbf{r} + (1/c^2)(\mathbf{p_1} \times \mathbf{p_2} \times \mathbf{r})) \qquad (19)$$

Next, we factor out the particle masses from the momentum term

$$\mathbf{F} = K*(c/R)^2(m_1 m_2 \mathbf{r} + (1/c^2)(m_1 m_2 \mathbf{v_1} \mathbf{x} \mathbf{v_2} \mathbf{x} \mathbf{r})) \tag{20}$$

$$\mathbf{F} = K*(c/R)^2(m_1 m_2)(\mathbf{r} + (1/c^2)(\mathbf{v_1} \mathbf{x} \mathbf{v_2} \mathbf{x} \mathbf{r})) \tag{21}$$

where **r** is the unit vector **R**/R.

Since our two masses form a closed, conservative system, we can 'normalize' our force by dividing by the total energy of the system; $E_{TOT} = m_1 + m_2$.

$$\mathbf{F}/E_{TOT} = K*(c/R)^2 \mu (\mathbf{r} + (1/c^2)(\mathbf{v_1} \mathbf{x} \mathbf{v_2} \mathbf{x} \mathbf{r})) \tag{22}$$

where $\mu(\mathbf{R}, d\mathbf{R}/dt) = m_1 m_2/(m_1 + m_2)$ is the reduced mass of the two body system.

In order to investigate the vector cross product term, we will assume our two masses (no longer necessarily electrons) are equal and orbiting one another.

Then we can write

$$\mathbf{F}/E_{TOT} = K*(c/R)^2 \mu (1 - (v^2/c^2)) \tag{23}$$

Seriously! And

$$\mathbf{F}/E_{TOT} = K*(c/R)^2 \mu - K*(\mu v^2/R^2) \tag{24}$$

We will call the second term in equation (24), the *coriolis* force.

If we recast our force equation into polar coordinates and allow $m_1 \neq m_2$ (i.e. for the study of planetary motion; Kepler's equations), we will pick up the usual *centrifugal* force term, in addition to our new *coriolis* force term.

$$\mathbf{F}/E_{TOT} = K*(c/R)^2 \mu - K*(\mu v^2/R^2) - K*(l^2/\mu R^3) \tag{25}$$

Finally, there will be a force term corresponding to the interaction of the spin/angular momentum (σ, **l**) of one object with the 'magnetic force vector' of the other object; F_{spin}.

(Remember, in our model, the relativistic mass of an object is due to the *total* relativistic motion of the mass; including spin!)

So, the total Lorentz force, for cosmology for example, will consist of four terms;

$$F_{universal} \sim F_{central} + F_{coriolis} + F_{centrifugal} + F_{spin} \tag{26}$$

In our generalized Lorentz force, the coriolis force is not an artifact of the choice of a particular reference frame, but arises from the absolute relative motion of the two bodies.

In our model, the coriolis force is *real*, because all forces are velocity dependent due to the fact that particle mass is velocity dependent; $m = m(\mathbf{r},\mathbf{v})$.

When we speak of the relativistic mass in our model, this includes *all relative velocities*, and not only linear velocities approaching the speed of light.

Similarly, the centrifugal force is now also a real velocity dependent force and is not due to "space time" disturbances as in the general theory of relativity.

Equation (25) should be 'easy' to generalize to an N body system.

The Lorentz torque:

The 'Lorentz torque' arises from the interaction of the intrinsic angular momentum (quantum spin or classical moment of inertia) of one body with the magnetic field vector of the second body, yielding the force 'F_{SPIN}' introduced above.

Unfortunately, this 'force' is actually a torque, which seems to complicate adding it to our Lorentz force equation. However, if we consider the potential energy instead of forces, the problem goes away!

$$E_{SPIN}/(E_{TOT}) = (\mathbf{I_1} \cdot \mathbf{B_2})/(m1+m2) + (\mathbf{I_2} \cdot \mathbf{B_1})/(m1+m2)$$

$$E_{universal} \sim E_{central} + E_{coriolis} + E_{centrifugal} + E_{spin}$$

Cosmology:

The general theory of relativity is no longer thought to describe the gravitational interaction of matter. Hence, many of the current problems in cosmology, which imply the inadequacy of the current theory, need to be reevaluated in the light of our new theory.

Large scale neutral matter interactions are described by the cosmological Lorentz force

$$F = m^*F\_g + m^*(v \times B\_g) \qquad (27)$$

It is important to note, that the mass of the objects involved is the *total relativistic mass*, which includes the mass due to rotation.

Angular motion is acceleration. It is also translationally invariant and thus considered to be absolute motion. Thus when considering the interaction between two galaxies, for example, one must include the relativistic masses of the individual spinning bodies, as well as the contribution to the total relativistic mass of the galaxy due to the bodies rotating about the center of the galaxy!

A body on the edge of a rotating galaxy will have a relativistic mass relative to the center, modifying the central force on the body and its angular velocity.

No more dark matter!

We can also now see why planets bulge. Matter on the equator of a planet is more massive than matter at the poles.

Also, there is now the prospect for repulsion between two bodies moving in opposite directions. In addition, a spinning galaxy now creates a 'gravitational magnetic field' and can be assigned a series of magnetic moments.

Obviously, if equation (27) is correct, then all the derivations already done in the study of electrodynamics can be applied directly to cosmology!

Our (*independently*) derived) theory is <u>very similar</u> to Gravitoelectromagnetism (GEM).

The only thing the GEM equations (analogs of the Maxwell equations) are missing is the mass current term, **J_m**, in the equation for electromagnetic induction, which we introduced in "On Parity and Isospin".

Red shift:

The gravitational contribution to the redshift of light from a distant galaxy is no longer attributed to the stretching of space. Instead, the shift in frequency is due to the gravitational work done on the photon by the mass of the emitting galaxy. This view will yield new relative velocities for the various galaxies, as well as, perhaps, new mass estimates.

Black holes:

In our new model, ordinary matter is composed of electrons, positrons and antineutrinos.

When the mass of a body reaches some critical point and the pressure becomes great enough that the force of gravity overwhelms the electrostatic repulsion responsible for the structure of the protons and neutrons, the constituent electrons and positrons will annihilate producing photons. Of course, we will also have the residual electron antineutrinos.

So, we would have a mix of bosons and fermions. It seems the photons would all fall into the same lowest possible frequency energy state, perhaps defined by the diameter of the black hole, and form a Bose-Einstein condensate. The antineutrinos might either escape, or remain gravitationally trapped, thus giving the black hole it's physical extent (and extra mass) due to the 'Pauli exclusion principle' keeping the neutrinos apart.

In our model, black holes consist of photons and electron antineutrinos, as well as whatever other detritus happens to get sucked in.

A black hole can be considered to be an 'infinite' spherical potential well of radius R. We imagine that the gravitational potential is constant inside the black hole

$$V(r) = G*M/R \; ; \qquad ; r < R \qquad (28)$$

and falls off in the usual way for $r > R$,

$$V(r) = G*M/(R+r) \qquad ; r > R \qquad (29)$$

The energy levels of the antineutrinos inside can be calculated using well known techniques. Since the potential energy of the well is not actually infinite, there will be the possibility for the high energy neutrinos to tunnel their way out!

For the photons, we don't like the idea of 'plane wave' solutions (as they would be reflecting off of a "potential barrier"), and so propose instead, a collection of di-photon bound states.

It's all quite speculative at the moment!

The eternal universe:

We predict the picture of the universe that will emerge after a reevaluation of the current cosmological data, will be one of an eternal, infinite, and *static* universe.

From lowly elementary particle decays to exploding stars, our universe is constantly being reseeded with new materials to build new stars, new solar systems, new galaxies.

On time and temperature:

Entropy.

All together now …   boring and confusing!

Fortunately, there is no room for entropy in our cosmological model.

There is no equation of the universe. Nothing so grandiose!

Our universe is an infinite, static, eternal, yet 'closed system'.

Our model of the universe can be considered as 'closed' in two ways; 1) in the sense that there are no "external" influences!, and 2) any system of interest (electron-proton, earth-moon, earth-sun, galaxy-galaxy, etc.) *can be*, and are routinely, appropriately and effectively isolated, both experimentally and theoretically, to produce sensible predictions and results.

For cosmological studies, one would choose an appropriately 'isolated' system (e.g. all the galaxies in the local cluster) and apply the analogues of equations (25) and (26).

Space is space. Space is the same here, there, and everywhere. From the space between atoms to the space between the stars,

It is all the same space.

Quantum mechanical interactions:

In our new model, the photon facilitates all interactions between particles, just as in quantum electrodynamics, except that in the universal model the photon couples to the relativistic mass of a particle as well as the electric charge.

We shall construct our new quantum mechanical coupling charge in strict analogy with the classical coupling charge of equation (15).

The photon then couples to a particle's 'total coupling charge', tcc, which is defined (with the sign convention fixed by equation (15), rightly or wrongly!) to be

$$tcc = (m/a)*(b - e') \qquad (30)$$

where m is the particle's relativistic mass and e' is the charge to mass ratio as before, and

$$a = (4*pi*epsilon\_0*hbar*c)^{1/2} \qquad (31)$$

and

$$b = (4*pi*epsilon\_0*G)^{1/2} \qquad (32)$$

The tcc is designed to yield

$$alpha = e^2/(4*pi*epsilon\_0*hbar*c) \qquad (33)$$

for purely electromagnetic (and non-relativistic) interactions, and

$$alpha\_G == m\_e^2*G/(hbar*c) \qquad (34)$$

for gravitational interactions. (There will be an 'unfortunate' cross term between alpha and alpha_G in charged particle interactions, which we will discuss later on.) Anyway ...

$$tcc = (m)(G^{1/2} - (e/m_e)(1/4\pi\varepsilon)^{1/2})(1/h^{bar}c)^{1/2} \qquad (35)$$

For quantum mechanical calculations, the propagator for the interaction between any two particles is then

$$f(q) = (tcc\_1)*(tcc\_2)/q^2 \qquad (36)$$

where q is the four momentum of the exchanged photon.

We can see from equations (35) and (36), that the energy scale dependence of an interaction is now determined by the mass of the interacting particles, rather than by the mass of the propagator as described in the standard model. In our new model, the propagator is always massless.

Finally, the photon also has a coupling charge, m_g, given by

$$m_g = \hbar \nu / c^2 \qquad (37)$$

where nu is the frequency of the photon and c is the speed of light.

Nonrelativistic quantum mechanical gravitational interactions:

We began our enquiries in reference (1) by considering a gravitationally bound state of two identical electron neutrinos; neutrinium. We assumed our two neutrinos would be bound by the classical Newtonian gravitational potential and thus based our neutrinium model on the analogous, and well known, positronium system bound by the Coulomb potential.

The Bohr radius was derived (missing an all important factor of G, here remedied!) to be

$$R_{neutrinium} = 2(4\pi\hbar^2)/(G m^3) \qquad (38)$$

where m is the neutrino mass. We then compared this result to that expected from general relativity,

$$R = 2 h^2/(G m) \qquad (39)$$

Noting the discrepancy of a factor of $1/m^2$ between the two results, we had to conclude that the general theory of relativity was incorrect (1,2).

It had a good run!

The strong force:

We remind the reader that we have defined the gravitational "rest charge" of a particle to be equivalent to its rest mass.

We boldly extend our gravitational model to the strong force and assume that a particle has a "rest color charge" analogous to our gravitational rest mass charge. A particle's color charge would then increase with its momentum. This model would account for quark confinement and asymptotic freedom.

At this point, it actually seems unnecessary to keep the color charge at all, so we declare

rest color charge == rest mass charge

If this is the case, then the strong force is just relativistic gravity.

Since we have no more need for the color charge, it seems reasonable to dismiss the quarks as well, and assume all partons are actually leptons.

In this model we have,

proton   = (e+,e-,e+)

neutron = (e+,e-,v_e^bar)

electron =   e-

and there is no more mystery of the missing antimatter. Bonus!

In the universal model, the strong force is merely the relativistic gravitational interaction balanced by the electromagnetic interaction, plus an additional interaction due to our surprising new 'cross-term'!

In our new strong force model, the linear force term arises from the mass of a particle and is multiplicative; rather than including an additional force term which is linear in the separation R.

Also, there is a natural attenuation of this force as particles approach relativistic velocities; hence, it can never become infinite.

As an added bonus, there are no more fractionally charged quarks, summing over colors, etc.

Let's look at the total coupling between two electrons.

For the sake of simplicity, and for illustrative purposes, we will begin with the non-relativistic case, where m*e' ==> e.

The total coupling, $(tcc\_e)^2$, is then, from equation (30)

$$(tcc\_e)^2 = (1/a^2)*(e^2 - 2*e*m*b + m^2*b^2) \qquad (40)$$

Using equations (31) - (34), we have

$$(tcc\_e)^2 = alpha - 2*e*m*b/a^2 + alpha\_G \qquad (41)$$

We shall call the cross term, alpha_strong, where

$$alpha\_strong == (G/(4*pi*epsilon\_0))^{½}*(2*e*m/hbar*c) \qquad (42)$$

Remembering that equation (42) is non relativistic, we can replace e with m*e' to obtain

$$alpha\_STRONG = (G/(4*pi*epsilon\_0))^{½}*(2*e'*m^2/hbar*c) \qquad (43)$$

We see from equation (41) that our new strong force term acts as a counterbalancing force, changing sign for different combinations of electrons and positrons.

The coupling alpha_strong would be appropriate for describing interactions between nucleons (protons and neutrons), while alpha_STRONG would be the choice for parton interactions within the nucleons.

In our model of the proton and neutron (and all baryons), the leptons are highly relativistic, so no simple hand waving model is possible. On the other hand, we do not have to worry about, or demand, antisymmetric parton wave functions even for identical partons within a baryon.

The running of alpha:

In the standard model, the electromagnetic coupling strength is expressed in terms of the electric charge, e;

$$\alpha = e^2/4\pi\varepsilon \hbar c \qquad (a)$$

In our model, the 'electric charge' looks like this

$$e \rightarrow (m^*e/m_e) \qquad (b)$$

and alpha becomes

$$\alpha = m^2(e/m_e)^2/4\pi\varepsilon \hbar c \qquad (c)$$

$$\alpha = (m_e^2/1 - v^2/c^2)(e/m_e)^2/4\pi\varepsilon \hbar c \qquad (d)$$

$$\alpha = (1/1 - v^2/c^2) e^2/4\pi\varepsilon \hbar c \qquad (e)$$

$$\alpha = \alpha_0 (1 + (v/c)^2 + (v/c)^4 + \ldots) \qquad (f)$$

In our running of alpha, the velocity squared replaces the four-momentum transfer, Q^2, of the standard model and there is no need to introduce an arbitrary cut-off mass.

In the standard model, the higher order Feynman diagrams for a scattering process represent the perturbative expansion of a *single integral*, in terms of the fixed coupling constant alpha, leading to the *running* of alpha, charge screening, renormalization, etc.

These higher order diagrams represent vacuum polarization loops, multiple photon exchanges, and all the other pathologies of quantum field theory.

In our model, there *is* only one integral as there *is only one* virtual photon involved in the interaction or scattering. There are no fields, there is no vacuum, and there is no charge screening.

In our model, e/m_rest is a fixed coupling constant and the relativistic mass of a particle is the only variable, so our perturbative expansion of alpha is in terms of $(v/c)^2$.

Rather than imagining a "high Q^2 probe" penetrating a virtual electron cloud and thus increasing the effective coupling charge, we have an increased coupling charge because the particles involved in the scatter are highly relativistic.

The weak interaction:

In the universal model, weak interactions always involve the neutrino, and thus are weak due to the tiny neutrino mass.

One of the conclusions of "A Quantum Mechanical Theory of Gravitational Interactions" was that a particle's 'weak charge' was equivalent to its mass, and that this tiny coupling charge was enough to explain the 'weakness' of the weak interaction.

Even so, we did not address the existence of the W and Z bosons out of convenience and expediency (and with no clear ideas on the matter at the time!). In this section, we will examine whether the W and Z are really necessary for weak interactions, and whether we can explain them away!

Let's begin historically, with beta decay; whereby a neutron turns into a proton;

$$n \longrightarrow p + e- + v\_e\textasciicircum bar \qquad (44)$$

In the standard model, beta decay occurs when a down quark emits a massive W- boson to become an up quark. The W- particle then decays into an electron and an electron antineutrino.

$$(u,d,d) \longrightarrow (u,u,d) + e- + v\_e\textasciicircum bar \qquad (45)$$

This process is highly suppressed due the massiveness of the W and not due to the nature of the weak charge.

In our new model, where partons are leptons, beta decay now looks like this

$$(e+,e-,v\_e\textasciicircum bar) \longrightarrow (e+,e-,e+) + e- + v\_e\textasciicircum bar \qquad (46)$$

Here, the antineutrino emits a photon which then decays into an electron and a positron. In terms of the Feynman diagram, the coupling charge at the antineutrino vertex is the neutrino mass (which 'suppresses' the interaction).

$$alpha\_weak \;==\; m\_v\textasciicircum 2*G/(hbar*c) \qquad (47)$$

The coupling charge at the electron vertex is the tcc of the electron.

This view of the weak interaction does not require a charged quantum mechanical force field. It is already a 'stretch' to imagine fields possessing energy and momentum, much less mass and electric charge!

As a second example, let's look at muon decay,

mu+ ---> v_mu + e+ + v_e   (48)

Again, in the standard model, the muon emits a charged W particle turning into a muon neutrino. The W+ then decays into a positron and an electron neutrino.

In addition to requiring a virtual field to carry the mass and electric charge away (as in beta decay), it just seems unnatural that a particle should change into another particle, particularly in such an *ad hoc* way.

For muon decay, we imagine the muon emitting a muon neutrino just as in the standard model, but rather than coupling to a virtual W, the muon couples to a virtual, massless, charged lepton which then emits an electron antineutrino, thus becoming a real electron.

This process is shown in Figure 1.

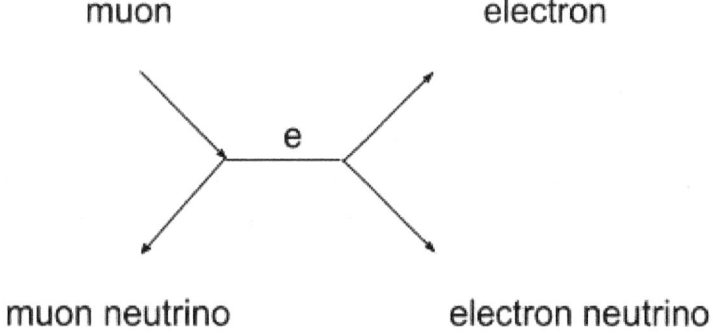

FIGURE 1: Muon decay. The propagator is most likely a generic virtual lepton, e/mu/tau.

In our model, the muon 'sheds' energy and spin in the form of its neutrino. The energy and spin shed ensures that the ensuing virtual lepton propagator has spin = 0, and is *massless*. In our model, all 'propagators' are massless.

This requirement places constraints on the energy and momentum of the initial and final states in muon (and tau) decays, and should help to explain the energy hierarchy of the three particle families. (A homework assignment!)

So, we have accounted for the weak interaction, with a standard model type approach, but without having to invoke charged, massive virtual exchange bosons or fields.

This new weak theory is essentially QED with mass as the coupling charge. Beauty, eh?

"Hold on", you might say, "haven't the W and Z been observed in particle collisions?"

In our new theory, these particles represent temporary resonant bound gravitational states of the colliding leptons. (For W production in ppbar collisions, it is equally likely to imagine the event as deep inelastic scattering between an electron and a neutrino.)

Our conclusion is the electroweak theory gives the correct energy scale for the unification of electromagnetism and the 'weak interaction', but an incorrect picture of the mechanism.

As an example, let's consider Z production in electron positron collisions.

In the standard model, the electron and positron annihilate producing a Z (or technically Z/gamma!) which can then decay into a pair of leptons or a pair of quarks which then decay into 'jets' of particles. So, our new model must explain this hadronization.

In the universal model, the electron and positron can annihilate to a photon or form a temporary, gravitationally bound, mesonic state.

Once the electron and positron are bound gravitationally, gravity effectively starts to behave like QCD in terms of confinement. The particles can only 'break apart' by forming gravitational bonds with neighboring particles.

Of course, this is only a rough analogy. We will need a slightly more rigorous picture of the gravitational hadronization and the gravitational strong forces for a complete model.

More homework!

The electroweak bosons:

The massive bosons of the standard electroweak theory are no longer thought to be 'field quanta', but are now considered to be resonant bound states of the two fundamental leptons.

$W^+ = (e^+, \nu_e)$

$W^- = (e^-, \bar{\nu_e})$

$Z = (e^+, e^-)$

The Z looks to be an (higher) excited state of positronium; the first excited state would be assigned to the pion. Or perhaps, the Z could be made up muons instead, or an admixture of all three (known) charged leptons!

In this model, the newly observed Higgs' boson would then be a bound state of an electron neutrino and an electron antineutrino.

$H = (\nu_e, \bar{\nu_e})$

The vacuum:

In the universal model, the Higgs boson is no longer required to provide the particles with mass, and indeed, the universe is no longer permeated by the Higgs field (or the W field, or the Z field, or the quark and gluon fields, etc.!).

In the standard model, the Higgs boson was thought to arise from a symmetry breaking of the energy of the vacuum. Now, this apparent feature of the 'physical vacuum' no longer seems to be necessary.

In our new model, the vacuum has no energy and no symmetry (or asymmetry for that matter!) and is not a teeming cauldron of subatomic particles.

In our model, fields are not real and hence there is no quantum mechanical vacuum.

Regardless, even in quantum field theory, when performing the 'second quantization' of particle fields in terms of the simple harmonic oscillator, one has the option of retaining the *constants of integration* (thus *creating* the quantum vacuum!), *or* setting them to zero.

Such a matter of taste should not be the basis for any model of reality.

The Dirac equation:

Why is the normalization for a Dirac spinor $1/(E + m)^{1/2}$?

This is quite unsettling since the normalization is greater than the total energy of the particle by an additional amount corresponding to the rest mass of the particle.

This cannot be right!

The Dirac equation for a 'free electron' is

$$H \psi = (\alpha \cdot p + \beta m_0) \psi \qquad (49)$$

where the Hamiltonian H, is the total energy of the electron, and m_0 is the rest mass. The Hamiltonian operator is

$$H = i \, \partial/\partial t \qquad (50)$$

In our model, the mass-energy, $E = mc^2$, is the particle's coupling charge, and completely determines the particle's behavior. So the rest mass term in the Dirac equation seems, somehow, 'redundant', unnecessary, and *unwanted*.

Let's define the kinetic energy operator (T = m - m_0) to be

$$T = H - \beta m_0 \qquad (51)$$

We still don't like, or want, this extra, constant, additive, and *global* charge mucking up our beautiful equation, so (and I think you see what's coming ...) we make a '*global gauge transformation*' and the problem is gone!

The Dirac equation for a *massive* electron is now

$$T == H \psi = (\alpha \cdot p) \psi \qquad (52)$$

and the spinors for the left handed and right handed solutions are now *decoupled*.

In our model, the Dirac equation does not have positive and negative energy solutions. (A 'free particle' cannot have negative energy!) Instead, we have spin left (electrons) and spin right (positrons) solutions. There is no right handed electron in our model or in nature!

There is no positron sea, and no particles traveling backwards in time.

If we add interactions to the mix, the general form of the Dirac equation will be

$$(T - V) \psi == L \psi = 0 \tag{53}$$

where, of course, L is the Lagrangian operator.

Our 'gauge invariant' solution of the free particle Dirac equation would look something like;

$$\psi = \exp(-i^*m_0) \exp(-iE^*t) \exp(i\mathbf{p} \cdot \mathbf{x}) \tag{54}$$

Replacing E(t) with the relativistic mass m(t), gives

$$\psi = \exp(-i^*m_0) \exp(-im^*t) \exp(i\mathbf{p} \cdot \mathbf{x}) \tag{55}$$

The first factor in equation (55) is a global phase factor representing the invariant rest mass-energy/charge of the particle. This is an extra and annoying charge which is already included in the second factor of the equation; i.e the total mass-energy/charge, m(t), which we take to be our locally variable 'phase factor' *and* the locally variable interaction charge of the particle.

However, our theory is not really a gauge theory, and there seems to no formal way (e.g. by a series expansion) to have the global phase factor in equation (55) cancel the rest mass energy term in the Dirac equation.

So ... we take our lead from the standard model and throw the rest mass term away!

However, we needn't 'recover' the rest mass from an interaction with the Higgs field as in the standard model. In our model the contribution of the rest mass energy is already accounted for in the total relativistic mass energy of the particle, which is the particle's interaction charge.

To introduce interactions into the Dirac equation, we modify the standard replacement

$$p_\mu \rightarrow p_\mu + eA_\mu \tag{56}$$

to include the gravitational interaction, and to account for the relativistic mass as the fundamental interaction charge, and obtain

$$p_\mu \rightarrow p_\mu - i^*\hbar (G^{1/2} - e/m_{rest}) A_\mu \, \partial/\partial t \tag{57}$$

This substitution should then yield the Universal Interaction Lagrangian

$$L_{interaction} = -i\hbar(G^{1/2} - e/m_{rest})(\bar{\psi}\gamma^\mu A_\mu \partial\psi/\partial t) \qquad (58)$$

We interpret the component A_μ as representing a *virtual* photon, and thus we have no need for the antisymmetric tensor $F^{\mu\nu}$, and the corresponding E and B fields, to facilitate the interaction or to carry energy and momentum. Remember, in our model, all the energy and momentum in an interaction is carried by the particles!

The virtual photon is already coupled to, or 'tethered' between, two mass-charge currents, only one of which is indicated in equation (58). The second current could be a similar leptonic current or a real photon current!

The wave function of the virtual photon propagator would be

$$A_\mu = \varepsilon(\mu)\exp(-i((m_1-m_2)/\hbar)t) \qquad (59)$$

where m_1 and m_2 are the relativistic masses of the interacting particles.

In our model, we expect to treat the real and virtual photons as separate particles, each with their own 'wave function'.

Gauge theory:

We can now understand the origin of gauge invariance in QED.

The variable mass charge involved in an interaction is compensated for by a time dependent 'phase factor' in the wave function; $\exp(-iE^*t)$.

Technically, our theory is *not* a gauge theory, although it does exploit and explain the observed gauge invariance of the electromagnetic interaction (another ingenuous discovery of the standard model, to be sure!).

In our model, the resolution of the 'gauge invariance issue' does not involve, or allow, the capricious variation of local charge with corresponding and compensating potential fields. (This is a feature never truly exploited, or properly explained, or *necessary* in QED ....)

Instead, in the universal model, we have a locally varying charge because the mass-energy is the coupling charge of a particle, and this varies during an interaction.

In conclusion, since the standard model explanation (and implementation) of the required 'gauge invariance' observed in QED is now thought to be incorrect, the generalization of gauge theory to explain the weak and strong interactions (thus yielding the W, the Z, and eight gluons) is probably also incorrect.

Of course, we have already removed the W, the Z, and gluons and quarks from our model. However, the more corroborating arguments for a new theory, the better.

Gauge invariance:

The universal model takes advantage of, and explains, both global and local gauge invariance.

Global gauge invariance allows us to remove the rest mass of a particle from our equations, because it is a global, constant, and therefore uninteresting, contribution to the particle's variable mass-energy charge.

Local gauge invariance is necessary because a particle's mass-energy charge varies during an interaction.

The Born approximation:

In the Born approximation, one assumes the scattering interaction between two particles can be described as the exchange of one virtual photon. This photon is considered to be the most energetic of all the photons exchanged during the interaction, and the contributions from "higher order" photons are found to be negligible in cross section calculations.

One of the mysteries of the standard model is why this first order approximation from the perturbative expansion of the interaction in terms of the coupling constant alpha, yields an exact result.

In our model, *there is only one* virtual photon involved in the interaction between the two particles, and its behavior is described by equation (59).

Hard scattering occurs as the virtual photon exchanged between the two particles becomes highly energetic. The frequency of a "hard scatter" in an experiment depends on the incident particle beam densities and individual particle impact parameters (and not on some random, or probabilistic, emission of a high Q^2 virtual photon from the 'target'!)

In the single particle exchange picture of the standard model, one particle is the emitter, and the other is the receiver. This is an excellent mathematical representation. Physically, it is lacking as the exchange is not 'symmetric', and it is hard to understand how such an exchange could lead to an attractive force.

In reality, the two particles have been interacting and exchanging virtual photons long before they were even collimated into colliding beams! However, we needn't go that far back in time.

The point is, in their final approach down the straightaway, the two particles are interacting furiously. The particles are always interacting; before, during, and after the 'scatter'.

Energy and momentum are always being exchanged between the two particles. There is no emitter and no receiver.

In our model, the q^2 value of the 'single exchanged photon' represents the total q^2 value given by the integral of equation (59).

For every action there is an equal and opposite reaction.

The classical propagator:

We now revisit our idea of creating a 'propagator' from our new universal Lorentz force.

The complete, 'classical', relativistic, Lorentz force between two identical electrons is

$$\mathbf{F} = (G/c^2 - (e/m_e)^2(\mu/4\pi))(c/R)^2(m_1 m_2 \mathbf{r} + (1/c^2)(\mathbf{p_1} x \mathbf{p_2} x \mathbf{r}))$$

where, of course, $\mathbf{F_1} = -\mathbf{F_2}$.

We shall start our study slowly, and consider only the static, or Coulomb potential, as one usually does.

$$\mathbf{F}_{1,2} = K^*(c^2/(\mathbf{r_1} - \mathbf{r_2})^2 (m_1(\mathbf{r_1}, \mathbf{v_1}))(m_2(\mathbf{r_2}, \mathbf{v_2}))$$

The intriguing factor of c^2/R^2 (the result of some factoring), already looks 'propagator like', with units of $1/t^2$.

We could look at the work

$$W_{1,2} = -W_{2,1} \Rightarrow W_{1,2} + W_{2,1} = 0$$

or even the action. I believe the key to this problem is converting the integrals such that we are integrating over the masses, dm, of the two particles; the limits of integration being the initial and final relativistic masses of the two particles.

We expect the solution will look a lot like equation (59)

The building blocks of matter:

In the universal model there are two fundamental fermions, the electron and the electron neutrino; and one fundamental boson, the photon. From these building blocks, all ordinary matter is formed

The electron neutrino is thought to be the fundamental unit of mass. It is considered to be a 'true' point particle of spin ½. From a consideration of the relative strengths of the electromagnetic and weak interactions, the rest mass of the neutrino has previously been derived (1) to be

$$m\_v = m\_e/e \tag{60}$$

where m_e is the electron mass and e is the magnitude of the electric charge.

The electron neutrino is also believed, from arguments of beauty and symmetry (13), to have a magnetic moment given by

$$mu\_v = (m)(hbar/2*m\_v) \tag{61}$$

where m is the relativistic mass and m_v is the rest mass of the neutrino.

In our model, the motion of mass is considered to be the ultimate cause of all magnetic and electrical forces or fields. The 'spinning mass' of the (point) neutrino gives rise to its magnetic moment.

In addition, the electron neutrino is thought to 'spin to the left', while the antineutrino spins to the 'right'.

In the universal model, leptons now have two spin degrees of freedom! (?)

All leptons have a 'fixed spin' of left or right, as well as an 'interaction spin' which can point up or down. Particles spin left and anti-particles spin right. Particles maintain their defining direction of spin (left or right), while the spin ½ component can flip up or down during an interaction.

So, in our model the neutrino has an antiparticle.   Neutrinos spin to the left and antineutrinos spin to the right. Similarly, electrons spin to the left and positrons spin to the right.

This explains why we only observe left handed neutrinos in interactions with the electron.

Matter is left handed. Antimatter is right handed. We live in a sinister universe!

Since we consider the electron neutrino to be the fundamental unit of mass, we shall now invert equation (1) to obtain a formula for the electron mass in terms of the neutrino mass and the value of the electric charge

$$m_e = e * m_\nu \qquad (62)$$

The neutrino mass can be taken from measurement or, ideally, derived from first principles!

We hypothesize that the electron neutrino is the "self-gravitationally" bound state of 'one quantum of action', h. We will revisit this idea in a later section.

We can now construct the universal model of the electron magnetic moment in strict analogy with the neutrino magnetic moment of equation (61), and in accordance with our new generalized Lorentz force, and the nature of our new coupling charge.

$$\mu_e = (m)(e/m_e)(\hbar/2) \qquad (63)$$

using equation (62)

$$\mu_e = (m)(1/m_\nu)(\hbar/2) \qquad (64)$$

and equation (61)

$$\mu_e = (1/m_\nu)(\mu_\nu) \qquad (65)$$

In this model, according to equations (62) and (63), the electron is 'just' the neutrino with an electric charge. The electron has increased in mass-energy and coupling charge by a factor of e relative to the neutrino.

We hypothesize that the electron is an 'excited' neutrino and, that the electron's magnetic moment *and* its electric charge are caused by the motion of spinning mass.

We have called this idea 'quantum mechanical electromagnetic induction'. Here, opposite directions of spin would generate opposite electric charges (i.e. either electrons or positrons). The generated electromotive force (Joules/Coulomb) of the spinning mass would manifest itself as an electric point charge.

We assume the rest mass and the spin of the electron are constant, so this would imply that electromotive force is quantized like all other physical quantities and processes.

Also, the neutrino would have no charge in this model because it is considered a true point particle without 'a middle'.

In our new model, the electron is 'not quite' a point particle, although it is considered to be an electric point charge! The (extended) rotating mass of the electron is thought to give rise to the point electric charge by "quantum mechanical electromagnetic induction" (3). This electrical energy seems to contribute to the electron rest mass, and the value of e seems to be tied to the quantization of the mass of charged particles as demonstrated by equation (62).

Similar relations are assumed to hold for the muon and the muon neutrino as well as for the tau and the tau neutrino. Hence, we can easily predict the masses of the muon neutrino and the tau neutrino from the measured masses of the muon and the tau.

The particle model:

Particles are not waves, or wave packets, or fields, or the disturbance of fields. Particles do not pop out of the vacuum. Particles are not wave functions. A particle is not a Dirac delta function. Particles do not interact with fields, either to exert or experience force, nor to acquire mass. Fields are not real.

These are mathematical models of particles; descriptions of particles; descriptions of the behaviour of particles; descriptions of the interaction of particles

They are not particles. They are not even an *approximation* of particles.

Nowadays, it has become the fashion to state that all particles are actually 'field-particle hybrids', that they don't really spin, and probably don't even have an intrinsic mass!

Why are people so enamored of such counterintuitive, illogical, unphysical, and irrational ideas?

I believe it is a mark of modern sophistication; as if due to our (solely) physical and technological advances, we are now somehow smarter or wiser than Isaac Newton or Immanuel Kant!

As if.

The actual claim seems to be that the math says so. However, we now demand that the math be in accord with the world, rather than the world with the math (i.e. kludged up Lagrangians, infinities, singularities, divergences, ever multiplying forces and fields, etc.)

The world is mathematical, but the world is *not* mathematics.

We shall now try to construct a more sensible phenomenological model of the fundamental particles; their world, and ours.

In the universal model, half integral spin is *the* defining, fundamental, and irreducible representation of any massive inertial particle.

In the sinister universe, the manifestation of half integral spin is the very *essence* of mass and inertia.

A particle of half integral spin can always be assigned a definite position in time and space. No other particle may occupy this same position in space at the same time.

The neutrino:

The neutrino is a tiny blob of rotating matter.

The neutrino may be considered to be a point particle, just as the earth, a galaxy, or even a cluster of galaxies may be considered point particles.

We assume the neutrino has no (particle) substructure, but this does *not* mean that the neutrino does not have extension. Who knows?

The electron neutrino *is* a tiny blob of mass of spin ½, spinning to the left.
The electron neutrino has a magnetic moment proportional to its mass

The electron antineutrino *is* a tiny blob of mass of spin ½, spinning to the right.
The electron antineutrino has a magnetic moment proportional to its mass
(and equal and opposite to the magnetic moment of the electron neutrino).

Two electron neutrinos moving under mutual gravitational attraction *cannot* collide.

Two electron neutrinos *cannot* occupy the same point at the same time.

Two electron neutrinos will ultimately repel one another due to the interaction of their magnetic moments.

An electron neutrino and an electron antineutrino, moving under mutual gravitational attraction, *will* collide, and annihilate one another.

We picture this annihilation as the cancellation or interference of the spins of the two particles, yielding a massive, yet *inertialess* particle, with spin 0; the 'virtual photon'.

The photon:

The photon serves at the pleasure of the leptons; either as the virtual gauge boson mediating their interactions, or as a real particle; for the shedding of energy and momentum of any lepton undergoing and/or *resisting* acceleration.

In our example of neutrino antineutrino annihilation, the subsequent virtual photon may couple to any lepton antilepton pair, *or*, to two real photons. The photon is a self-coupling gauge boson, because it carries gravitational charge (mass-energy).

A bound electron can only make an energy transition (e.g. in the hydrogen atom) if the subsequent change in angular momentum of the system is +/- 1. This is because the electron spin has to flip to produce a real photon of spin=1 as it *accelerates*.

The photon is produced (generated, released) by the electron. It does not pop out of the vacuum. It is not an excitation of a field.

The photon is energy, momentum, and spin 'shed' by the electron

The photon carries just enough quantities to perform its job, which is transferring energy and momentum (and spin) between interacting fermions!

We know the energy of a photon is proportional to its frequency. However, when a photon loses or gains energy, it does not change in 'size', and its spin remains a constant.

The only physical attribute the photon possesses is spin. This spin can be positive or negative relative to the transverse direction of the photon.

We hypothesize that the frequency, nu, of a free photon is the rate at which the photon spin flips, changing polarization harmonically as it barrels toward its target at the speed of light. This 'rate of spin flip' would be what would give the photon a particular energy and momentum.

Our new model of the world is a mechanical model; albeit without the gears, wheels, and pulleys, that Maxwell and Faraday imagined.

Fundamental particles are spinning mass-energy. Half integral spin is the origin of inertial mass. The mass-energy of a particle consists entirely of rotational kinetic energy. This rotational kinetic energy is due to the particle's spin, *and* spin precession about the direction of the linear momentum of the particle.

The linear momentum of a particle (or more specifically, impulse) is due to a particle's helicity flipping harmonically at the characteristic frequency that defines the energy and 'wavelength' of the particle.

This is true for the photon, as we have argued just above, as well as for the electron, as we shall demonstrate in the next section.

Particle in a box:

Let's begin with the classic example of the particle in a one dimensional box.

A particle (e.g. an electron) in a box bounces back and forth between the walls. The particle follows a well defined path and physically traverses *all points* lying between the walls of the box.

The current interpretation of quantum mechanics posits there are points where the electron never is; points the electron does not pass through; points along the trajectory where the electron essentially ceases to exist!

Of course, these points are the nodes where the quantum mechanical wave function of the particle (a sine wave) passes through zero.

In our new model, these nodes represent points where the electron is actually *physically unable* to interact because it is undergoing a 'spin flip', or a 'change in helicity'.

We now understand the spinor nature of the electron, and why it's spin is 'double valued'.

As the electron travels along, its spin precesses once about the direction of travel (it's 'proper' or measured helicity, say) and then the spin flips, going around once the other way; ad infinitum.

At the instant of the helicity flip, the electron is not able to interact with 'real' photons.. These points correspond to the nodes of the quantum mechanical wave function.

We can now interpret the probabilities generated from squaring the wave function to indicate where a particle is actually able to physically interact and *be measured*, rather than where the particle is.

In addition, like a swimmer doing laps in a pool, the boundary conditions of our box dictate that the electron must perform a helicity flip when it bounces off the walls; hence the wave function must yield a zero expectation value at the boundaries of the box.

This also explains the nature of the Bohr orbits in the hydrogen atom, and why they must correspond to integral values of the electron wavelength.

In our new model, the electron is an inertial (i.e. spin ½) blob of mass-energy/charge, spinning to the left. The axis of the electron spin and the principal axis of rotational inertia of the electron are aligned or 'projected' along the direction of motion of the electron and precess about this direction with a frequency proportional to the total mass-energy of the electron.

This precession frequency corresponds to the de Broglie wavelength of the electron;

$$v = E/h == m/h \qquad (66)$$

$$\lambda = h/p = h/mv \qquad (67)$$

with wavenumber

$$k = 2\pi/\lambda \qquad (68)$$

and so it takes 2π radians for electron spin to precess once about the direction of motion.

However, since the electron is a 'spinor', it takes 4π radians, or two revolutions of the spin vector, for the precessing *angular momentum vector* to return to its original value and helicity.

A free electron of fixed helicity, executes a 'polarization flip' every 2π radians, performing a 'complete revolution' every 4π radians. This spin flipping is what gives a particle its oomph!

Particle in a box:

```
                                                       ^
                                                        \
                                                         \
            |-------------------<-------------------------\|
                                                      ^
                                                       \
                                                        \
            |-------------------<------------------------\---|


            |-------------------<-----------------------/---------|
                                                       /
                                                      /
                                                     /

   node     |----------<----------------.------------------------|

                                              ^
                                             /
                                            /
            |----------<-------/------------------------------|

                                      ^
                                     /
                                    /
            |-----<---/--------------------------------------|




            |---------\--------------------------->------------|
                       \
                        \
                         \

   artist's conception
```

To reiterate, an electron in a box bounces back and forth between the walls. The electron follows a well defined path and physically traverses *all points* lying between the walls of the box. The behavior of the electron spin vector is illustrated on the previous page.

The wave function for an electron in a box, indicates where, when, and how 'efficiently', the electron is able to interact.

Figure 2 shows the probability distribution for the state n = 2.

The electron spin precesses around the direction of motion, performing a helicity flip every 2 pi radians, and completing one complete 'revolution' every 4 pi radians.

At the point x = L/2, the electron is undergoing a flip in helicity, where its spin is effectively zero, and hence it cannot interact with our experimental probe (e.g. a photon).

Similarly, at the walls of the box, the electron must perform a helicity flip as it bounces, exchanging a virtual photon with an electron in the wall of the box.

At the point x = L/4, the electron helicity is in 'full bloom' and it can interact with a real photon.

The electron oscillates harmonically between the ability to engage in real and 'virtual' interactions

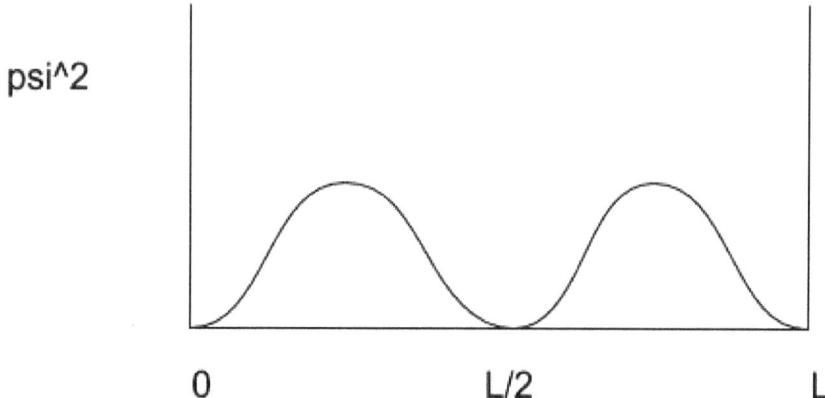

Figure 2: Probability distribution for a particle in a box; n = 2.

Hidden Variables:

We have discovered the hidden variables of quantum mechanics.

They are called spin, mass, and charge!

More specifically, the hidden variables are; the *phase* of the particle wavefunction, the spinor nature of the electron, and the double valued state functions of the electron.

In the standard model, the phase of the wave function, and the spinor nature of the electron, are considered curiosities and dismissed, since they do not contribute to the expectation value of any observable when one squares the wave function.

In our model, the phase of the particle wave function is *real,* and has real, observable, and measurable effects.

In our model, subatomic particle interactions are completely deterministic, and there exist many physical and 'timing' constraints on these interactions, since the electron, as a spinor, can only interact at certain times, and in certain ways; alternatingly interacting with real and virtual photons, as its helicity oscillates back and forth, as it travels a *well defined* path.

We speak of interactions as characterized by the exchange of real or 'virtual' photons.

The irony is that the interactions that occur via 'virtual' photon exchange may actually be considered the 'more real'.

Real photons emitted by a particle may interact with a subsequent particle, but this is completely accidental. Only human beings employ real photons to influence other particles.

For this reason, we prefer to characterize interactions as extrinsic vs. intrinsic, external vs. internal, or accidental vs. incidental.

They are all real interactions.

The single slit experiment:

In our model, the characteristic 'interference' pattern observed on the screen or photographic plate in a single slit diffraction experiment, is not due to two separate electron waves arriving at the screen in or out of phase and reinforcing or canceling each other out.

Instead, *each separate electron* arrives at the screen 'in or out of phase' for being able to interact with the photographic plate.

An individual electron can only darken the photographic plate if it arrives with the proper phase. Just like our particle in a box, the oscillating helicity of the electron has to be in the process of 'flipping polarization' when it arrives at the screen (i.e. the electron must have helicity ~zero to interact with the surface of the plate).

Electrons that do not successfully interact with the surface of the plate will carry on until they *do* find something to interact with.

We suggest a single slit experiment where the photographic plate is replaced with several layers of silicon, analogous to tracking detectors employed in high energy scattering experiments.

Our prediction is, of course, that the electrons which fail to contribute to the pretty pattern on the top layer, will show up in subsequent layers; perhaps displaying secondary pretty patterns.

The double slit experiment

We can employ similar arguments to explain double slit interference patterns, although in reality the situation is much more complicated.

In the single slit experiment, it does not really matter what happens before the slit. The slit is essentially the source of the electrons and the electrons have a characteristic spread determined by the width of the slit and the uncertainty principle.

For the double slit experiment, the environment of the electrons before the two slits *does* matter.

Although an electron can only pass through one slit or the other, it feels the 'potential' of both slits.

Remember, our 'free' electrons are actually exchanging virtual photons with *everything*. On their approach to the two slits, they are interacting with the wall housing the two slits, and they are interacting with the screen (which is their final destination) by virtual photon exchange via the two slits.

How this could actually be cast and analyzed in terms of potentials is unclear!

We suggest a similar double slit experiment, where the the two slits are replaced by two similarly sized and spaced metal plates, set to a negatiive retarding potential; thus, scattering the incoming electrons back toward the source and an appropriately placed screen.

The wave function:

The wave function is longer mysterious or spooky.

You can no more predict the position of an electron bouncing around in a box than you can predict the position of an ideal gas molecule bouncing around in a box.

The wave function is a mathematical function invented by human beings. It has nothing to do with the particle 'itself'. The wave function is *not* the particle and does not 'collapse' when one makes a measurement anymore than a momentum vector describing the particle would collapse.

The wave function is a probabilistic determination of what energy, momentum, and position values we can expect to measure for a particle *we cannot see*.

The wave function is not a real, physical thing. It has nothing whatsoever to do with the particle; either physically, or 'metaphysically'.

The wave function does collapse.

It is a completely meaningless notion and a *non problem*.

Quantization:

Ultimately, quantization boils down to the satisfaction of boundary conditions.

Virtual photons can only attach to/interact with an electron when the electron has an effective helicity of zero (during its polarization flip) because virtual photons are spin zero.

This fixes the 'frequency' of the virtual photon coupling two interacting electrons, since virtual interactions, or the exchange of energy and momentum, can only occur when both electrons have helicity ~= 0.

When an electron reaches 'peak helicity', it is able to interact with (i.e. absorb and emit) real photons the most readily and the most efficiently.

Magnetic moments:

Previously, we introduced the universal model generalization for the formula of the electron magnetic moment;

$$\mu_e = (e/m_e)(m)(\hbar/2m_e) \tag{69}$$

This formula can be interpreted as

$$\mu_e = (\text{coupling constant}) \ast (\text{mass}) \ast (\text{angular momentum per unit mass})$$

So, the electron is one unit of angular momentum per unit mass per unit volume of space; or one unit of inertial, half integral, angular momentum per unit mass.

$$e \Rightarrow h/4\pi m_e == (\hbar/2)/m_e == L/m_e \tag{70}$$

Inserting the formula for the relativistic mass into equation (5) we get

$$\mu_e = (e\ast\hbar/2m_e)(1/(1 - v^2/c^2)^{1/2}) \tag{71}$$

We then make the usual series expansion to obtain

$$\mu_e = (e\ast\hbar/2m_e)(1 + \tfrac{1}{2} v^2/c^2 + \tfrac{3}{8} v^4 c^4 + ...) \tag{72}$$

and find the magnetic moment of the electron increases with velocity, *as expected*.

The electron is essentially the vector **L**/m_e. This vector precesses about the axis of the direction of motion with a frequency; $\nu = E/h = m/h$. As the speed of the electron increases, the frequency of precession increases, and the mass of the electron increases.

The surprising 'spinor' nature of the electron is due to the angular momentum vector flipping helicity/polarization every 2π radians.

Even though the mass and magnetic moment of the electron increase with the velocity, the electron angular momentum is always $\hbar/2$.

We finally know what has been waving all this time!

The waving of the wave equation/wave function represents the periodic precession of the electron spin about the direction of travel of the electron!

N.B. The wave equation describes *many* physical phenomena that don't really wave.

Spinoring:

We keep saying the "helicity or polarization" of the electron 'flips' every 2π radians.

Technically, of course, this terminology is incorrect, because even though there *is* 'flipping' going on, the helicity of the electron never changes!

Instead, we shall say the electron is 'spinor-ing', as illustrated in Figure 3.

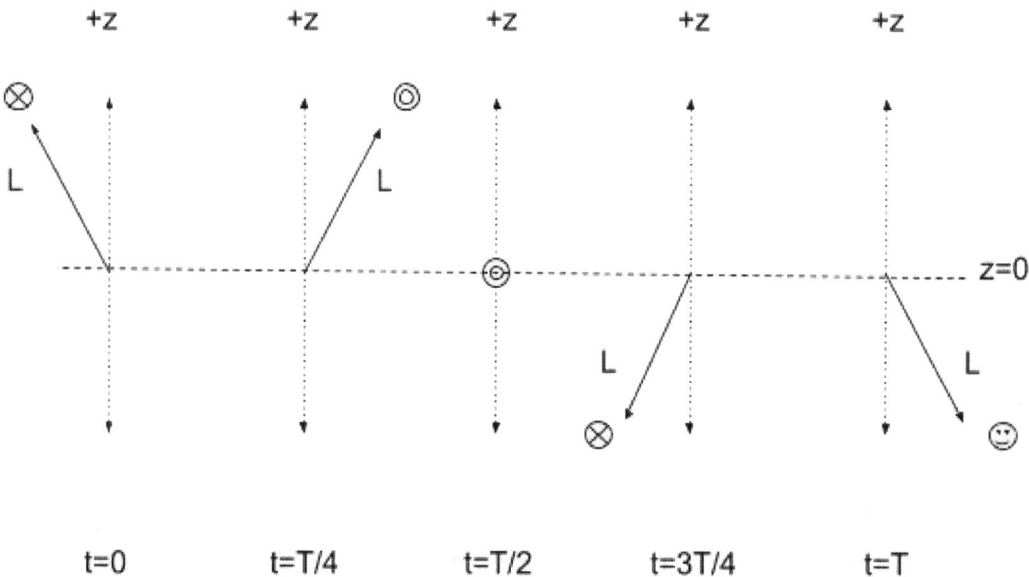

Figure 3: At rest with an electron traveling the the z-direction. The spin angular momentum vector 'precesses' about the direction of motion, tracing out a closed, three dimensional figure eight (a string!). The x symbol represents motion into the page. The dot symbol represents motion out of the page. At time T/2, we see the the angular momentum is *perpendicular* to the direction of travel. (This is when the electron engages in 'virtual' interactions.)

We also can see, that although the angular momentum vector is 'spinor-ing', the polarization, or helicity, of the electron does *not* change, and is constant!

The neutrino magnetic moment is

$$mu\_v = (m)(hbar/2m_\nu) \qquad (73)$$

Inserting the formula for the relativistic mass of the neutrino into equation (73) we see

$$mu\_v = (hbar/2)(1 + \tfrac{1}{2} v^2/c^2 + \tfrac{3}{8} v^4c^4 + ...) \qquad (74)$$

And, we must conclude that *the neutrino is one inertial quantum of action* as we proposed in "On Matter, Mass, and Motion". We can represent the neutrino symbolically, as we did for the electron in equation (70), and write

$$nu\_e \Rightarrow hbar/2 == L \qquad (75)$$

Actually, equation (74) shows that *there is only one neutrino*. The three neutrinos differ only by their **velocity.**

We can now see that neutrino mixing is due to elastic collisions of the neutrino with matter, decelerating the neutrino, and changing its flavor.   </blink>

The tau lepton is the most massive of the three charged leptons, so when it 'decays', it emits the highest velocity, and hence, most massive of the the "three" neutrinos, etc.

The tau is shedding energy and spin, because what else is there?

We can also derive the magnetic moment of the neutrino using the hand waving arguments usually assumed for the electron magnetic moment.

$$mu = (mass\ current)(area) = (m/t)(A) = m(L/2m) = hbar/2 \qquad (76)$$

Finally, we reach way back to Physics 101 and recall the formula for angular momentum

$$L = mvr \qquad (77)$$

The angular momentum of the neutrino is h and we assume it spins with angular velocity c.

$$h = (m\_nu)(r\_nu)c \qquad (78)$$

Now, we can solve for the *Compton radius of the neutrino*; the smallest probable distance.

$$r\_nu = h/(m\_nu)*c \qquad (79)$$

The photon, neutrino, and electron:

The photon is essentially a perpetual motion machine! The photon polarization oscillates harmonically at a frequency proportional to its energy, nu = E/h. This is an unorthodox picture of the photon polarization, but we note it satisfies the orthogonality condition on the photon wave vectors; $\mathbf{k} \cdot \varepsilon = 0$. The photon is corpuscular! (Newton: 2, Everybody else : 0)

The photon is one 'free', massive, but inertialess, unit of angular momentum; L = h.

The neutrino is one 'bound', massive, unit of angular momentum *per unit space*; L = hbar/2.

The electron is one 'bound' unit of angular momentum per unit space *per unit mass*; L = hbar/2.

Particle interactions:

Take a look at the angular momentum vector of the electron at time, t=0, in Figure 1. Here, we say the electron helicity is in 'full bloom' and it is able to absorb a real photon (*if* its polarization is also 'blooming'), increasing the rate of precession of the electron angular momentum vector, and hence the electron mass.

Consider the case of partial transmission and partial reflection of light from a thin sheet of glass. Photons in 'full bloom' will be transmitted. Photons with polarization ~0 will be reflected.

Similar arguments can be made for electron tunneling. If an electron arrives at a potential barrier 'out of phase for reflection', *and,* the electron 'wavelength' is comparable to the 'height' of the potential barrier, the electron will 'tunnel' through!

Quantum field theory:

In our model, particles are not created and destroyed. Instead, particles absorb, emit, merge with, and *shed* one another.

For example, an electron and a positron do not 'annihilate', producing a virtual photon.

The electron and positron have equal and opposite spins, ½. They merge to become a photon of spin 0. The photon then splits into the particle antiparticle pair demanded by the situation.

In our model, particle interactions are a continuous flux and flow of energy and momentum, *flowing only one way*; futureward. Particles interact by exchanging units of *angular momentum*.

Particle families:

In our new model, we have replaced the quarks with leptons. However, there is not a one-to-one correspondence between the two species, and 'the mapping' will not necessarily be the same between the different particle groups of the standard model.

In our model, the first generation of particle families is now imagined as follows:

proton = (u,u,d) ⇒ (e+,e-,e+)

neutron = (u,d,d) ⇒ (e+,e-,$\bar{v_e}$)

electron ⇒ electron

electron neutrino ⇒ electron neutrino

In the standard model, the second generation consists of the strange quark, the charm quark, the muon and the muon neutrino.

The charm quark and anti-charm quark can form temporary bound states in high energy collisions ("charmonium" a.k.a the J/Psi meson) which then decay into a muon antimuon pair or an electron positron pair. The strange quark has yet to be seen in such a bound state. Hence, for the second generation, we make the following substitutions;

charm quark ⇒ muon

strange quark ⇒ muon antineutrino

Similarly, for the third generation, we have:

bottom quark ⇒ tau meson

top quark ⇒ tau antineutrino

We still assume the relationships concerning mass, charge and spin derived for the electron and the electron neutrino hold for the higher generations as well.

Why is there more than one particle family?

What dictates the energy differences between the three particle families?

We hypothesize that when generating high energy leptons, at some point it becomes more expedient for 'Nature' to generate a muon rather than an electron with a velocity v/c ~= 1.

This appears evident from equation (74) and our new theory of the neutrino.

If true, where is this point? Is there a hard and fast rule governing when to chose the higher family particle, or is there some probabilistic indeterminacy involved?

In our model, the muon 'sheds' energy and spin in the form of its neutrino. The energy and spin shed ensures that the resulting virtual lepton propagator has spin = 0, and is *massless*.

This requirement places constraints on the energy and momentum of the initial and final states in muon (and tau) decays, and should help to explain the choice of the final state lepton from the three particle families in a particular decay, as well as some of the current mysteries surrounding lepton universality.

We can solve for this threshold, by assuming the muon rest mass is equivalent to the largest *allowable* relativistic mass of the electron.

$$m\_mu = m\_e/(1 - v^2/c^2)^{1/2} \tag{79}$$

$$(v^2/c^2)_{THRESHOLD} = 1 - (m\_e/m\_mu)^2 \tag{80}$$

A similar calculation will result in the velocity threshold between the mass of the muon and the mass of the tau.

Group theory:

Remarkably (and necessarily, perhaps), the universal model retains the SU(3)SU(2)U(1) group structure of the standard model.

However, these groups do not generate 'exchange quanta'. In addition, the color charge is no longer the basis for the group SU(3). Instead, we propose the three charged leptons; the electron, the muon, and the tau.

The group SU(3) reflects an exact symmetry of the universal model, operating on both the charged lepton basis

e = (1,0,0) ; muon = (0,1,0) ; tau = (0,0,1)

and the neutrino basis

nu_e = (1,0,0) ; nu_muon = (0,1,0) ; nu_tau = (0,0,1)

The neutrinos have the same mass hierarchy (~1,100,1000) as the charged leptons.

The eigenvalues, q, of the charge operator are degenerate for the charged lepton basis, but the eigenvalues q/m are non-degenerate, as are the masses themselves.

We can use the step up and step down operators (I+ and I-), composed of the Lambda_i matrices of the standard model, to transform between the particles in each of the two independent bases. Neutrino mixing, anyone?

We can also borrow the formalism of flavor SU(3) by making the substitutions

up → electron

down → muon

strange → tau

and generate all the pseudoscalar mesons, for example;

pi^0 = 1/(2)^½ (e e^bar - mu mu^bar)

Finally, we can generate particles by mixing the charged lepton and neutral lepton bases, and the neutral lepton bases, using the Lambda_i.

The fundamental lepton:

We call our fundamental lepton, the tetrahedron, in honor of Plato; the Man, the Myth, and the Legend!

Our tetrahedron may, or may not, have four sides; but we will assume it has three principle axes of rotational inertia.

In our theory, *all inertia* is rotational inertia. Trying to accelerate an elementary particle is analogous to trying to walk around a high school physics lab while holding a spinning bicycle tire.

In addition, the mass-energy of a particle is due to the rotational kinetic energy about its axis of rotation (at rest), as well as an additional rotational kinetic energy which arises due to the the rate of spin precession of the rest mass around the direction of travel for a particle in motion.

We imagine the three principle axes of inertia of the neutral tetrahedron to correspond to the three neutrino masses, and the three principle axes of the charged tetrahedron to the masses of the electron, the muon, and the tau.

More specifically, in light of equation (70), there is one charged tetrahedron with three principle axes of rotational inertia (L/m_e, L/m_{mu}, L/m_{tau} ; $L = \hbar/2$) and one neutrino to serve them all!

There are now three electromagnetic coupling constants; e/m_e, e/m_{mu}, e/m_{tau}.

The step up and step down operators composed of the Gell-Mann matrices of SU(3), are actually geometrical rotations from one principle axis of inertia to another. When the tetrahedron flips from one axis to another, it sheds its energy and momentum in the form of its corresponding neutrino.

There are three particle families because there are three dimensions of space.

Similar rotations in "weak isospin" space transform the neutral tetrahedron into its charged counterpart.

What is this crazy isospin space? I don't know!

We *do* know that by using various combinations of the step up and step down operators of SU(2) and SU(3), plus the parity operator, we can write any quantum mechanical interaction current in terms of the 'fundamental' neutrino neutral current.

Weak isospin:

Geometric rotations of the three principle axes of inertia of the tetrahedron about the axis of *spin*, generate the three lepton families, transforming one into another; all manifestations of a single particle.

These rotations correspond to the transformation matrices of the group SU(3) of the standard model.

Weak isospin rotations transform the neutral tetrahedron into the charged tetrahedron.

It seems these do not correspond to any real, physical, rotation.

We speak metaphorically of the electron as an excited neutrino; its electric charge a consequence of quantum mechanical electromagnetic induction (qemf).

However, qemf is quantized, so the electron cannot decay or collapse into a neutrino.

Charged leptons can 'decay' into one another, but the charged tetrahedron is conserved.

In conclusion, there are two fundamental leptons. One is neutral and one is electrically charged.

The leptonic table:

LEPTONS   ANTI-LEPTONS

electron	electron neutrino	PARITY ⇔	electron antineutrino	positron
⇐	CHARGE	MASS ↕	CHARGE	⇒
muon	muon neutrino	PARITY ⇔	muon antineutrino	anti-muon
⇐	CHARGE	MASS ↕	CHARGE	⇒
tau	tau neutrino	PARITY ⇔	tau antineutrino	anti-tau
⇔	weak isospin	mass isospin ↕	weak isospin	⇔

TABLE 1: The leptons and their interrelations.

Any lepton can be 'generated' from any other by the appropriate applications of the parity operator, the weak isospin operator, and our newly proposed 'mass isospin' operator.

The Heisenberg uncertainty principle:

    The Heisenberg uncertainty principle reflects the fact that our experimental probes perturb or destroy the microscopic system under investigation.

    It is *not* a ghostly feature of reality reflecting the murkiness of the physical properties of the particles and/or their *very* existence

The Pauli exclusion principle:

    Now that our most fundamental lepton, the electron neutrino, has a magnetic moment,

    ***and** we now have a completely mechanical model of subatomic particle interactions*,

    we can understand (as most people have long suspected) that the Pauli exclusion principle is due solely to the electromagnetic interaction between leptons -- which act like tiny little refrigerator magnets!

    There are no more mysterious quantum mechanisms.

The mechanical universe:

                                       Bicycle wheels and refrigerator magnets!

The hydrogen atom:

One of our conclusions from investigating the boundary conditions of an electron in a box was that analogous boundary conditions must apply to the electron in the hydrogen atom

We are bringing back closed Bohr orbits which correspond to integral values of the electron wavelength!

In addition, we assume the first Bohr orbit is *circular,* even though the angular momentum of the system has a value of zero.

N.B. We made this same assumption in our analysis of the neutrinium system which then allowed us to compare our results to those of general relativity.

In the ground state of the hydrogen atom, the spins of the proton and electron are aligned (+½, +½) for an angular momentum of +1. The orbit of the electron has an angular momentum of -1. The total angular momentum of the ground state is zero.

The magnetic moments of the proton and the electron are anti-aligned and repel one another keeping the electron from spiraling into the proton.

The stability of the hydrogen atom is now due to the physical necessity of closed electron orbits and the interaction of the magnetic moments of the proton and the electron.

Quantized orbits are now the result of ordinary physical interactions

O.K. -- *now*, there are no more mysterious quantum mechanisms!

The two pillars of twentieth century physics:

The two great pillars of twentieth century physics, general relativity and quantum mechanics, are incompatible and irreconcilable because they are both incorrect.

Conclusion:

Nature is mechanical and *dynamical*. Particles and their interactions can be completely described and explained in terms of fundamental, 'solid', and *real* units of matter, constantly in motion, and continually exchanging energy and momentum.

No more ghosties!

Isaac Newton --   best physicist ever!

Immanuel Kant -- most boring genius!

A quantum mechanical theory of everything:

photons, electrons, neutrinos; interacting synchronously

Books by Greg Feild:

the pentateuch

1. "A quantum mechanical theory of gravitational interactions"
   CreateSpace Independent Publishing, 8/29/2016

2. "Observations on the quantum mechanical nature of gravity"
   CreateSpace Independent Publishing, 10/8/2016

3. "On gravitation and electric charge"
   CreateSpace Independent Publishing, 10/29/2016

4. "On spin, mass, and charge"
   CreateSpace Independent Publishing, 11/29/2016

5. "On angular momentum, acceleration, and absolute motion"
   CreateSpace Independent Publishing, 1/1/2017

the exegeses

6. "The Sinister Universe"
   CreateSpace Independent Publishing, 3/1/2017

7. "On Parity and Isospin"
   CreateSpace Independent Publishing, 4/11/2017

8. "Reflections on the Sinister Universe"
   CreateSpace Independent Publishing, 5/12/2017

the hermeneutics

9. "On Current Physics"
   CreateSpace Independent Publishing, 6/11/2017

10. "A Critical Examination of Classical and Quantum Mechanical Waves"
    CreateSpace Independent Publishing, 6/18/2017

<u>the gospels</u>   :)

11. "On wave particle duality and the quantum of action"
    CreateSpace Independent Publishing, 7/6/2017

12. "On matter, mass, and motion"
    CreateSpace Independent Publishing, 9/14/2017

13. "On action and reaction"
    CreateSpace Independent Publishing, 9/24/2017

14. "A quantum mechanical theory of everything"
    CreateSpace Independent Publishing, 11/5/2017

<u>the compilations</u>

"The Universal Model of Our Sinister Universe: The First Ten Books"
CreateSpace Independent Publishing, 7/2/2017

"The Canons of the Sinister Universe:
The Last Four Books on the Universal Model of Our World"
CreateSpace Independent Publishing, 11/5/2017

Notes: :)

Eternal recurrence:

It's all the same -- day, man ...

-- Janis Joplin

On Interaction

Greg Feild

April 21, 2018

The history of natural philosophy is characterized by the interplay of two rival philosophies of time - one aiming at its "elimination" and the other based on the belief that it is fundamental and irreducible.

The basic objection to attempts to deduce the unidirectional nature of time from concepts such as entropy is that they are attempts to reduce a more fundamental concept to a less fundamental one.

-- G. J. Whitrow

Time is invention or it is nothing at all. But of time-invention physics can take no account ... Modern physics ... rests altogether on a substitution of time-length for time invention.

-- Henri Bergson

Apart from time there is no meaning for purpose, hope, fear, energy. If there be no historic process, then everything is what it is, namely, a mere fact. Life and motion are lost.

-- Alfred North Whitehead

Science in its effort to become more "rational" tends more and more to suppress variation in time.

-- Emile Meyerson

It cannot be too often emphasized that physics is concerned with the measurement of time, rather than with the essentially metaphysical question as to its nature ... We must not believe that physical theories can ultimately solve the metaphysical problems that time raises.

-- Mary F. Cleugh

Source (cherry picked from) : *Physics and the Ultimate Significance of Time*
                       Edited by David R. Griffin
                       State University of New York Press; 1986

Abstract:

This book examines and explores the
"Universal Model of Our Sinister Universe".

We will discuss the nature of our new model,
the general nature of physical models,
the nature of physics, the nature of science,
and consequently, and unavoidably, human nature.

:)

In order to seek truth it is necessary once in the course
of our life to doubt as far as possible all things.

-- Rene Descartes

About the author:

    Greg Feild writes for posterity,   :(
    but he hopes, and expects,
    to find readers today!

    He also designs his own book covers.

Coming soon:

    "On Rotation"

  It was a dark, and stormy night, in the sinister universe. Our hero, our *bulldog*, clutched their satchel ever closer; as inside was,

                                    the theory of everything.

Our *tenacious* hero, held, not *just,* the theory of everything, they carried

                                    the *quantum* theory of everything.

Thunder crashed like colliding stars . . .

Introduction:

    Old habits die hard.

    Old systems of belief, and old patterns of thought, seem to die even harder!

    Everyone is familiar with the story of the phlogiston.

    Everyone knows the story of the phlogiston, because everyone tells the story of the phlogiston to illustrate the historical errors in people's interpretation of physical process, and the subsequent 'paradigm shifts' in physics.

    The phlogiston, the crystal spheres, the electromagnetic ether, … ;
to this growing list of missteps and misunderstandings (i.e. *misinterpretations*),
we must add fields, the fabric of spacetime, and the phantasmagorical wave function.

    Our new <u>physical model</u> of the world, is not so much a paradigm shift, as a purging of all the current psychedelic, pseudo-scientific silliness sullying science;

    An exorcism of the painfully embarrassing, paranormal phenomena, plaguing (and continually accruing to!) physics since the infamous Copenhagen conference.

    We are *determined* to return *everyone* to the days of 'old school' physics, where, at the very least, physical theories dealt with physical phenomena, and scientific theories were confronted with experimental data.

    So, what is the phlogiston? No one remembers …

    Phlogiston.   It's a funny word!

                        Let's do physics!

                                            :)

The slippery slope:

When *did* physics begin the awful slide into inappropriate, unschooled speculation, and unfettered, unfounded, and unwarranted (and unbelievable!) supposition?

We believe, it all began the day physicists "boldly and triumphantly" dismissed the electromagnetic ether, and decided to let the electromagnetic fields "stand on their own".

Unfortunately, on that fateful day, they threw away the baby, but kept the bathwater !

(No actual babies were harmed in the employment of this metaphor.)

"We *now* know", the wave like appearance and properties of electromagnetic radiation can be accounted for by;

1) the periodic emission of photons from an oscillating source, and
2) the periodic oscillation of the polarization of the emitted photons.

Both these oscillations are of the same frequency; nu = E/h

It seems odd, on the discovery of the photon, that people did not immediately revisit the electromagnetic field theory.

Perhaps they had forgotten their initial discomfort when creating real electromagnetic fields existing independently in time and space.

Perhaps they were distracted, their incredulity dulled, by collapsing wave functions, ghostly, part-time particles, and the advent of the 'fabric' of time and space.

It must have been a heady, exciting, and *confusing* time, to be sure !

The result: "modern physics" is an *incoherent*, hot mess.

… and still confusing.

Just saying.

Symmetry:

The fundamental symmetries of the universal model, are the symmetries of space and time.

Space is flat, isotropic, homogeneous, smooth, continuous; Euclidean. Space is eternal and unchanging; from the past, through the present, and into the future. The space of today is the space of ten bazillion years ago. Space will be the same tomorrow.

Space is **where things happen**. TIme is **things happening**.

In *any* inertial reference frame (and we always choose the *universal* inertial reference frame), time is smooth, continuous, isotropic, homogeneous; Newtonian. Time is eternal, smooth, and unchanging; from the past, through the present, and into the future. The rate of flow of time of yesterday is the same as the rate of flow of time of ten bazillion years ago. Time will *continue* to flow in exactly the same way tomorrow.

These facts about time and space are what make the the derivation of the differential calculus, *and* the ability of the differential calculus to describe the rates of change of real world physical processes and dynamical systems, possible.

Now is the time for plain, pedantic, didactic, discourse. And *italics*, and **boldface type** !

The time of physics is (or should be) the quantification of ordinary time that people use to keep appointments, measure heart rates, the speed of race cars and sprinters, how long one can hold their breath, etc.

Time is *manifest* as a sequence of events; as a succession of moments.

Time *is* the ticking clock.

The time reversal invariance of our physical equations, represents the fact that the flow of time is a smooth, continuous, invariant, invariable, un-variable, relentless, inevitable, enduring, unchanging, and *constant* process.

This notion of time is the foundational basis of the calculus. Without this notion of time we *cannot* do physics.

If the equations of physics were *not* invariant under time reversal, this would imply that these equations were not employing our common, intuitive, and ordinary notions of time *correctly*.

Time is symmetric, from the past, through the present, and into the future, because it is a *constant* 'flow'. This fact *alone* allows us to cast our mathematical physical equations in terms of this important variable.

Time flows forward, **marking change**. *We are not allowed to assume* that time could/would/should run backwards, just because our mathematical equations seem to "allow" it.

This would be an incorrect *interpretation* of the math.

Time flows forward due to the conservation of energy and *momentum*.

It is really as simple as that. Really.

(The proverbial smashed vase will never reassemble itself.)

As for space:

One can only answer the questions; "How much space is over there?",
"How much space do you *see*?",

with the questions, "Between what and what and what?" "Bounded by what?"

One can measure a particular space by measuring the time taken for an object of known, fixed velocity (i.e. the photon), to travel from point A to point B, etc.

The fact that different observers, in different inertial reference frames, measure different time intervals for the same 'event', is analogous to the fact that objects weigh less on the moon than they do on the earth.

In the first instance, the nature, flow, and experience of Newtonian time is exactly the same in both reference frames, even though the observers may measure different relative times (*and positions*) for any given 'event'.

In the second instance, the mass of the object is constant and the relative difference in the measured weights is dependent on the choice of environment.

This is why, in the universal model, our absolute inertial reference frame is always taken to be the fixed background of space.

This is where we must put our fixed Cartesian coordinate systems; and *not* fixed at the center of a rotating body, and *never* at the center of a *massive* rotating body, in order to do physics correctly, and obtain the most accurate results possible from our calculations.

The second symmetry of the universal model is -

For every action there is an equal and opposite reaction.

This is true for the usual, central, radial force between two interacting bodies, *and* for the velocity dependent forces arising from the influence of the magnetic field vectors.

We know the magnetic forces "do no work", hence any angular motion induced in a first body due to the magnetic field of a second body, must be compensated for by an equal and opposite change in the angular motion of the second body due to the magnetic field of the first body.

Think of the spiraling, helical paths charged particles follow in an external magnetic field.

Subatomic particles behave the same way in the presence of each other's magnetic fields.

In addition, any torque exerted on the magnetic moment of the first particle, due to the magnetic field of the second particle, is equal and opposite the to the the torque on the second particle, due to the magnetic field of the first.

These equal and opposite torques may be considered as "central" between the two bodies.

The third symmetry of our model is -

The world is made up of equal parts matter and antimatter.
Matter spins to the left.   Antimatter spins to the right.

Our next "symmetry", concerns the microscopic and macroscopic 'regimes'.

In our model, the laws of physics are exactly the same in the subatomic realm and in the cosmic realm.

Time and space ~~behave~~ *are* exactly the same, everywhere, always.

The only difference between classical and quantum physics, is the nature of the _boundary conditions_ constraining the systems of interest.

The mantra of physics should continue to be;

Boundary conditions! Boundary conditions! Boundary conditions!

Hard to say and even harder to implement, but *very* important.

Our final "symmetry" is -

The gravitational and electromagnetic interactions behave exactly the same way In both microscopic and macroscopic systems; the only difference between the two (besides a giant scale factor), is the lack of a negative gravitational charge.

We whimsically call this 'broken' symmetry, e-Symmetry.

If this *is* a broken symmetry, then the culprit is, our 'newly discovered' phenomenon called quantum mechanical electromagnetic induction.

The same phenomenon that breaks the symmetry of *classical* electrodynamics.

(We now take a brief pause for people to tape their brains back together)

:)

OK !

The universal reference frame:

In our model, the force between two bodies orbiting one another, normalized by the total energy of the system, is (14)

$$F/E_{TOT} = K*(c/R)^2 \mu - K*(\mu v^2/R^2) - K*(l^2/\mu R^3) \tag{1}$$

We may write this symbolically as

$$F_{universal} = F_{central} + F_{coriolis} + F_{centrifugal} \tag{2}$$

These are all the "force terms". *However,* there is a contribution to the overall energy of the system, corresponding to the interaction of the spin/angular momentum (σ, **l**) of one object, with the 'magnetic force vector' of the other object.

$$E_{SPIN}/(E_{TOT}) = (\mathbf{l_1} \cdot \mathbf{B_2})/(m1+m2) + (\mathbf{l_2} \cdot \mathbf{B_1})/(m1+m2) \tag{3}$$

$$E_{universal} = E_{central} + E_{coriolis} + E_{centrifugal} + E_{spin} \tag{4}$$

These calculations must be performed in the universal reference frame as shown in Figure 1.

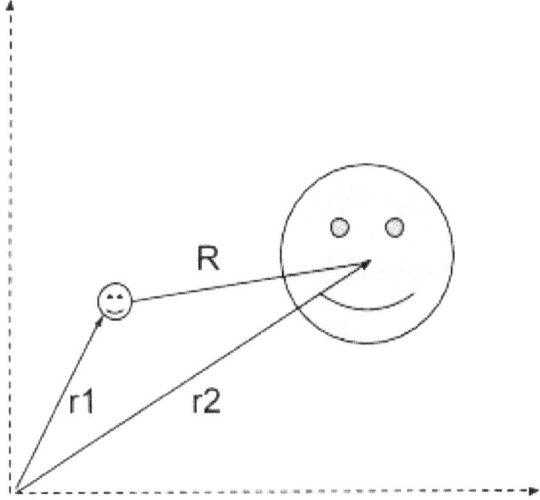

Figure 1: A cartesian coordinate system for the study of planetary motion. The coordinate system is in reference to the fixed background of space. The mutual force between the two bodies is a function of *all* relative velocities between the masses comprising the the two bodies.

Cosmology:

There are actually two more radial vectors necessary to completely describe the planetary motion depicted in Figure 1.

These are, of course, the radii of the orbiting bodies: r_A, r_B. Despite our previous admonitions, we *do* want to fix the coordinate systems for r_A and r_B at the center of the bodies A and B, respectively.

The relativistic mass of a celestial body includes a *considerable* contribution from its spin, and is dependent on the radius and rate of rotation of the body.

$$M = \Sigma_i \, m_i(v_i(r_i)) \qquad (5)$$

If we make the reasonable assumption that the angular velocities of the bodies are fixed during an interaction, we can make our first *approximation*, and perform the summation (integral) of equation (5) for the two individual bodies, before creating the reduced mass, μ, of equation (1).

We now have all the ingredients to do cosmology!

One chooses an 'isolated' system of interacting objects, and then applies the proper generalizations of equations (1) through (5).

All these interactions take place against the fixed 'backgrounds' of space and time.

"Sounds reasonable," you might say, "but how do you explain the gravitational redshift of light from a distant galaxy if space is a "fixed background" ?".

Gravitational redshift:

Since we no longer consider space to be a physical substance capable of expanding and contracting (it is now gone with the ether), we must assume that the gravitational redshift of light from a distant galaxy is due to the gravitational drag of the mass of the galaxy on the emitted photons. The photons are now considered to have a gravitational charge proportional to their energy since we believe mass and energy are equivalent.

The universe is no longer flying apart, but we can still use the gravitational redshift of light to measure the distance of a faraway galaxy by calculating the gravitational work done on a photon by the mass of the emitting galaxy.   (A fun math project!)

Forces and potentials:

One of the recurring questions of physics is; which concept is more fundamental, the concept of force, or the concept of potential (energy).

Since the concept of force is more reflective of everyday interactions that human beings have with objects, and considering the fact the concept of potential is derived from the concept of force, we choose force to be the more fundamental concept.

Note, we purposely and repeatedly employ the term concept to emphasize that neither forces or potentials are fundamentally real.

Forces and potentials are useful abstractions invented by human beings to help them understand and describe the world.

In our model, the concept of potential *energy*, has been **banished**.

However, just like the concept of force, we retain the more formal concept of potential for utilitarian purposes.

Rather than speak of the interactions between particles in terms of the balance between the kinetic energy and the potential energy of particles, we characterize interactions solely in terms of the changes in the kinetic energy and *relativistic mass* of the particles.

In our model, even the work done in raising an object a distance, d, above the surface of the earth,

$$W = mgd \qquad (6)$$

would actually be reflected and *formally* written as a change in the particle's relativistic mass-energy

$$m = m_{rest} c^2 + m_{rest} gd \qquad (7)$$

However, equation (6) is the more appropriate and useful choice for day to day applications.

:)

The interaction between (two) bodies is completely determined by -- the mass of the two bodies, the relative velocity of the two bodies, and the relative separation of the two bodies.

Or, more generally, and more succinctly,

The interaction between bodies is completely determined by the relative relativistic mass-energy of the bodies, and the distance between the bodies.

This is illustrated for the classic, classical, two body system by equation (1).

It all sounds so very simple, except the relativistic mass-energy of (even) a two body system is a complicated function of the several and various, relative *and* absolute, velocities describing the motion of the individual bodies.

As usual (or, as it was in the 'good old days'), the physics is easy, but the math is hard!

However, it seems much of the mathematics, and many of the mathematical tools, generated during the investigations into the general theory of relativity, could 'easily' be "mapped over" or "reworked", and applied in the pursuit and development of equation (1).

synergy

On action and reaction:

The "action" is a well defined quantity and concept in terms of;

1) "Minimizing the action" (i.e. minimizing the time integral of the Lagrangian, T - V, of a system of particles), yielding a quantity with units of (energy) × (time); (E • t).

2) Planck's "quantum of action" with units of E • t , representing one unit (the smallest possible) of 'interaction' that may be exchanged between two particles;

'interaction' *defined* as the exchange of (a certain amount of) energy
during (a certain amount of) time.

Everyone knows the units of the action, E • t, are equivalent to the units of angular momentum, kg m²/s.

Particles interact by exchanging units of action, or units of angular momentum.

Moving angular momentum around takes *work*, and, of course, everyone wants to minimize their amount of work!   "Work smarter, not harder!"

Nature does not like to do work. Nature's motto is: Do as little work as possible !

This is the true anthropic principle.   :)

Newton's third law is often colloquially expressed as "for every action there is an equal and opposite reaction".

Of course, Newton's third law is actually a statement about forces, $\mathbf{F}_2 = -\mathbf{F}_1$, and the term "action" is not well defined. Action seems to be used metaphorically to say that everything that happens to 'object one' during an interaction also happens to 'object two' in a compensating manner, ensuring the conservation of energy, momentum, angular momentum, etc., during the interaction.

This concept of action and reaction is a more general and more inclusive description of the interaction between two bodies, compared to demanding a simple balancing of the central forces, $\mathbf{F}_2 = -\mathbf{F}_1$

In the universal model, Newton's third law will be broadened to include forces as well as "actions".

Just as we 'generalized' the classical Lorentz force (subsuming Newton's universal law of gravitation in the process!), we now 'universalize' Newton's three laws of motion.

1) A body will not experience a change in **momentum**, unless acted on by a force, **F**, generated by a second body (or several other bodies).

2) $\mathbf{F} = d\mathbf{p}/dt = d(m\mathbf{v})/dt = m\, d\mathbf{v}/dt + \mathbf{v}\, dm/dt$ \hfill (8)

3) The two bodies experience equal and opposite changes in their respective *actions* during the *interaction* (which we have defined as the *exchange of units of action.*)

$$F_1 \times d_1 \times t = -F_2 \times d_2 \times t \quad (9)$$

and equal and opposite changes in angular momentum due to magnetic forces

$$\Delta L_1 = -\Delta L_2 \quad (10)$$

We now have equivalent and compatible definitions for "the action" in both the classical and quantum formulations of our universal model.

Newton's universal law of gravitation is *replaced* by the more general, and *properly normalized*, equation (1), and all calculations are done in the universal inertial reference frame of Figure 1.

The concept of 'minimizing the action' is already the mathematical bridge between classical and quantum physics, and remains so in the universal model, although we expect all Lagrangians to eventually be formulated strictly in terms of ΔT !

Otherwise, we like to avoid the use of the concept of potential energy for formal conceptions and definitions, and save it for the the practical applications of calculating the energy states of real world systems, where various approximations are necessary.

So, in our model, we envision the following substitutions; for the total energy

$$E_{TOT} = T + V = \text{const.} \quad \Rightarrow \quad m = T + m_{rest} \quad (11)$$

and for the Lagrangian

$$L = T - V \quad \Rightarrow \quad L = T - m_{rest} \quad (12)$$

where T includes *all* kinetic energy; linear, rotational, etc.

Mass *is* resistance to acceleration. If a mass is successfully accelerated, it tries to shed the extra energy, by emitting photons. Mass does not like to be accelerated!

Nature minimizes the action during an interaction, by keeping the 'relativistic' masses of the particles involved, as small as possible.

Nature achieves change (as it strives for an ever elusive stability) by doing the least amount of work possible, against all possible constraints, and within the boundary conditions fixing the closed system and defining the nature of the interaction.

To do the least amount of work, nature must keep acceleration to a minimum.

For any closed system, the sum of the individual energies divided by the total energy of the system, is equal to one; and we may write, symbolically;

$$(T + m_{rest}) / m = 1 \tag{13}$$

or more concretely,

$$((\Sigma_i T^i) + (\Sigma_i m^i_{rest})) / (\Sigma_i m^i) = 1 \tag{14}$$

In this formulation, we see all conservative interactions must satisfy one constraint

$$\Sigma_i (\Delta T_i) = 0 \tag{15}$$

During any interaction, or evolution of any closed dynamical system, the sum of the changes in the kinetic energy of the particles involved, must add to zero.

In our model, _only relative changes matter_; relative changes in velocity and kinetic energy.

Newtonian versus Hamiltonian dynamics:

The chicken or the egg?

We have already concluded, the concept of force precedes (or at least, preceded) the concept of energy, and, as such, must be considered the more fundamental concept, and the starting point of all investigations.

In the case of Newtonian versus Hamiltonian dynamics, we know which came first.

The questions people still like to ask are;

"Is one a more primary picture of particle dynamics?"

"Is one picture more reflective of reality ? --
of how the world 'really' functions, and operates, and behaves, 'beneath it all' ?"

Our answer is both ironic and somewhat paradoxical.

The Hamiltonian approach is more in accord with the spirit of our model, although the model is derived from the concept of force (of course!), and depends on the concept of force for many or most real world applications.

Still, we chose the Hamiltonian approach, as the better picture of the general behavior of particles, and their interactions, and as a better description of the dynamical evolution of a system of interacting particles.

In our model, particles do not feel force, or exert forces on one another, but rather, are constantly coupled, continually exchanging energy and momentum.

Let us say they are complementary.

On interaction:

In our model the fundamental particles are considered to be the physical incarnation, or manifestation, or *ensconcement*, of the fundamental unit of angular angular momentum; Planck's "quantum of action", h. This is the nature (or 'origin') of particle mass and spin.

This makes sense since intrinsic rotation, or spin, is translationally invariant, and apparently, compact, self-contained, and 'self-sustaining'. In addition, the intrinsic angular momentum of a particle (plus the precession of this angular momentum about the axis of the direction of motion) is the sole source, and origin, of inertial mass, relativistic mass, and gravitational mass.

We imagine the three fundamental particles as follows, as discussed in reference (14):

The photon is one 'free', massive, but inertialess, unit of angular momentum; $L = h$.

The neutrino is one 'bound', massive, unit of angular momentum *per unit space*; $L = \hbar/2$.

The electron is one 'bound' unit of angular momentum per unit space *per unit mass*; $L = \hbar/2$.

These three particles are the building blocks of matter in our model.

The leptons are the bricks, the 'virtual' photon is the mortar, and the 'real' photon is the wrecking ball!

The virtual photon is for 'internal bonding' only, in a 'closed' and 'conservative' system of interacting particles.

The real photon is for shedding unwanted energy and spin into the 'external' environment (thus, randomly wrecking things, or carrying the latest top forty tune to your car!).

Real photon *emission* is only necessary when our 'closed' system is perturbed by an external influence (e.g. in the electrification of hydrogen gas).

Either way, virtual or real,

There are three differences between the virtual photon and the real photon --

1) The virtual photon, spin 0, *transfers* energy and momentum *between* particles.
   The real photon, spin 1, *carries* <u>away</u> energy and momentum, *and* angular momentum (in units of h).

2) The virtual photon is imagined to be *a 'standing wave'*, of *varying frequency*, with the endpoints *secured* between two interacting, *moving* particles.
   The real photon is considered to be a free, 'traveling wave', of *fixed frequency*, *emitted* from an oscillating (and usually externally driven) source.

3) Virtual photons couple to the mass and spin of a particle.
   Real photons only couple to the spin of a particle;
   i.e. the *magnetic moment*.

The proper wave equation for a real, *literally free*, (no 'scare quotes' necessary!) photon is the Klein-Gordon equation. For simplicity, we assume travel along the x-axis, and denote the real photon wave function by the capital greek letter Γ.

$$\partial^2 \Gamma / \partial t^2 = \partial^2 \Gamma / \partial x^2 \qquad (16)$$

with solutions

$$E = +/- |\mathbf{p}| \qquad (17)$$

$$\Gamma_{LEFT} = \varepsilon \exp(-ip \cdot x) \qquad (18)$$

$$\Gamma_{RIGHT} = \varepsilon \exp(+ip \cdot x) \qquad (19)$$

We imagine the the positive energy solutions to correspond to spin left solutions, or those photons with negative (?) angular momentum, and the negative energy solutions to spin right photons, or those with positive angular momentum.

These assignments are based on our assumption that the electron is considered matter which spins to the left, while the positron is antimatter which spins to the right.

In our model, the photon is no longer its own antiparticle! There are now two distinct real photons. One couples to 'matter', the other, 'antimatter'.

A real photon is one unit of angular momentum, or action; h. Photons spin to the left and the projection of this angular momentum is along the direction of travel.

Anti-photons spin to the right, and the projection of the the angular momentum is antiparallel to the direction of travel or propagation.

As for the mode of photon propagation, we originally imagined a 'flipping' of the photon polarization as simple harmonic oscillation, with the photon polarization vector oscillating sinusoidally solely along the direction of travel.

This flipping action was introduced to explain how different photons can have different energies and momenta even though for every photon the angular momentum and velocity are fixed quantities; $L = h$, $v = c$. Only the frequency varies.

Upon further reflection, we realized this picture of photon propagation can not be physical and is not correct. We now believe the 'flipping' of the photon polarization vector to be more akin to the 'spinor-ing' of the electron which we introduced in "On Action and Reaction".

The 'corpusculating' of the photon is shown in Figure 2.

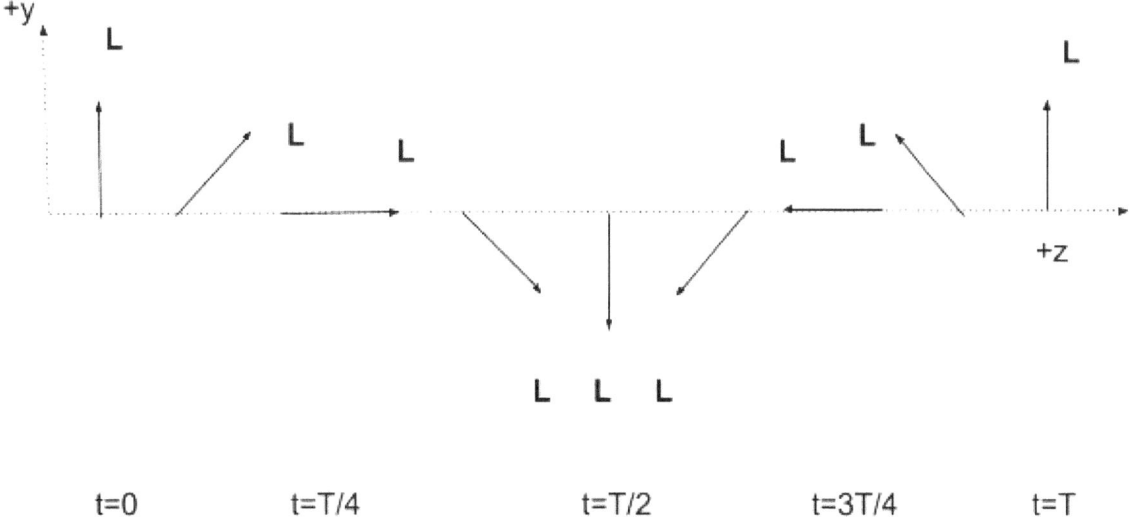

Figure 2: A 'plane polarized' photon traveling in the z - direction. (Individual real photons always spin to the left or right. One may imagine the momentum vector L spinning about the z-axis.) The photon rolls, or 'corpusculates', along the direction of travel projecting the spin angular momentum sinusoidally along the direction of propagation while maintaining a constant polarization and angular momentum; $L = h$. We could also say that the photon '**spirals**'.

The free photon satisfies the Klein-Gordon equation. In our interpretation, the positive and negative energy solutions correspond to positive and negative *helicity* solutions.

The free photon must also obey the Klein-Gordon probability current equation.

Again, instead of imagining positive and negative probability currents, we have two separate probability currents representing the two orthogonal, positive and negative, helicity solutions (i.e. left-handed and right-handed photons) of the KG equation.

We haven't quite worked out all the sign conventions describing the relations between the energy, angular momentum, helicity, and probability currents for photons and anti-photons, but the reader can easily see how the associations are to be made, at least in principle. (Homework!) We expect to present a more quantitative discussion of helicity and anti-particles in our next book "On Rotation".

A real, free photon always spins to the left or the right. Therefore, the wave function of a real, free photon must be a linear combination of the two orthogonal, transverse solutions of the KG equation. Does this suggest the wrong 'eigenbasis'; or perhaps that the Coulomb gauge is not the best choice for describing free photons?

To further investigate the behavior of the photon and the nature of photon interactions in our new model, let's look again the theory of photon "corpusculation" as depicted in Figure 2.

At time, t=0, we happen to catch the photon in the middle of a 'helicity flip'. At this moment, the photon angular momentum is entirely *transverse* to the direction of propagation. If the photon were to meet an electron at this particular moment, it could not interact because its angular momentum is 'unavailable'.

At time, t=T/4, the photon angular momentum is fully projected along the direction of travel, and the photon is in a position to transfer all its energy and momentum to an available and suitably oriented electron.

In our model, photons are self-coupling gauge bosons.
A photon and an anti-photon can annihilate producing a virtual photon;

(spin +1) plus (spin -1) = spin 0

In our theory, two properly oriented "two component" real photons can combine to form one "four component" virtual photon. It seems this particular virtual photon would have no longitudinal component . . . Just musing at the moment !

More generally, of course, a virtual photon connecting two electron has four components. This photon appears as a 'double spiral', like a double spiral staircase, with two components carrying energy and momentum in the **+r** direction, the other two in the **-r** direction.

Quantum mechanical electromagnetic induction:

In our model, not only do elementary particles really spin; they essentially *are* spin.

When it was first suggested that the electron might have a spin, it was decided that this was merely a 'quantum label' and could not be a *proper* classical spin, because the electromagnetic force was not sufficient to keep the electron from flying apart. This idea has not served physics well.

We now have the Magic of Quantum Mechanical Electromagnetic Induction to endow the electron with mass and electric charge, as well as stability and internal coherence.

Electromagnetic induction was the last classical electromagnetic phenomena to be discovered. It is not at all obvious, and cannot be derived from, or proved from, the other mathematical laws of electricity and magnetism.

Electromagnetic induction is the only classical phenomena that has not been observed in, or contemplated for, the quantum realm.

From arguments of beauty, symmetry, etc., we propose that electromagnetic induction is one of the fundamental laws of nature and must have a quantum mechanical counterpart.

Thus, quantum mechanical electromagnetic induction *just is a thing*, needing no further explanation or explication, and it gives rise to the electron and the electric charge.

Our speculations on quantum mechanical electromagnetic induction and the nature of the electron have been presented in several previous publications (references 3, 6, 7, 14).

We can summarize here by simply saying;

The electron is a puffed up neutrino.

The Dirac equation:

As previously stated in this paper and elsewhere, in our model all interaction is determined solely by the relative motion of massive bodies and can be completely characterized by the conservation of kinetic energy as demonstrated by equation (15).

In physics, we are not interested in the (essentially arbitrary) absolute 'total energy' of a particle or a system of particles, but rather in the *change* in energy over a prescribed period of time of interaction. In our model, any change in energy is a change in *kinetic energy*.

In this spirit, we will suggest a method to try and remove the rest mass from the Dirac Equation/Lagrangian in a 'rigorous' way.

The Dirac equation for a 'free electron' is

$$H \psi = (\alpha \cdot p + \beta m_0) \psi \qquad (20)$$

where the Hamiltonian H, is the total energy of the electron, and m_0 is the rest mass. The Hamiltonian operator is

$$H = i \, \partial/\partial t \qquad (21)$$

Let's define the kinetic energy operator ($T = m - m_0$) to be

$$T = H - \beta m_0 \qquad (22)$$

$$T = \alpha \cdot p \qquad (23)$$

In our model, the lagrangian is $L = \Delta T$, and

$$\Delta T = H \big|_{t=t2} - H \big|_{t=t1} \qquad (24)$$

Admittedly, this discussion is still a bit muddled ... maybe we'll have more later.

The real point we'd like to make about the Dirac equation today is that the primary operator and eigenvectors characterizing the Dirac spinor should be the observable $\sigma \cdot \mathbf{p}$ with the helicity eigenvectors +/- ½.

In this approach, just as in our model of the photon, the positive and negative 'energy' solutions of the Dirac equation are now interpreted as the (positive) energy solutions for the positive and negative eigenstates of the helicity equation (i.e. positrons and electrons).

To remind the reader of our definition of electrons and positrons, and the corresponding concept of electron propagation as 'spinoring', we reproduce here the diagram first introduced in "On Action and Reaction" as Figure 3.

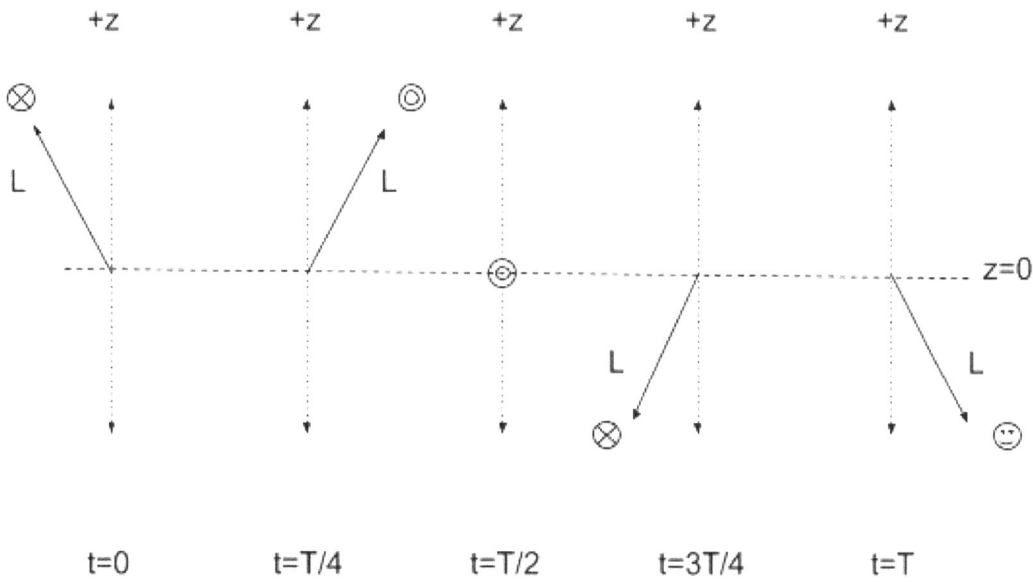

Figure 3: At rest with an electron traveling the the z-direction. The spin angular momentum vector 'precesses' about the direction of motion, tracing out a closed, three dimensional figure eight. The x symbol represents motion into the page. The dot symbol represents motion out of the page. At time T/2, we see the the angular momentum is *perpendicular* to the direction of travel. (This is when the electron engages in 'virtual' interactions.)

To represent a positron, simply swap the x symbols and the dot symbols.

In Figure 3, we can begin to see inklings of the uncertainty principle. Ideally, at time, t=0, the energy of the electron can be measured exactly. However, we cannot make instantaneous measurements as measurements always take some finite amount of time.

At the other extreme, at time, t=T/2, the electron is unable to interact with our apparatus at all!

In between, the fraction of energy available for interaction and measurement varies sinusoidally.

Und so weiter.

Wave particle duality:

When you want to study the 'wave like nature' of the photon or the electron, you must employ the particle wave function 'in real time' in order to include the contributions from all local and global phase factors influencing an interaction.

The particle phase, previously dismissed as a curiosity, actually determines the timing, place, and nature (i.e. 'real' versus 'virtual') of a particle's interaction with neighboring matter.

When studying the wave like nature of particle interactions, we must use the real part of the imaginary wave function, yielding trigonometric functions that describe the real time behavior of the particles, predicting when, where, and in what manner they will interact with our detector (the proverbial photographic plate).

When you want to study the particle like nature of the photon or the electron, you employ the square of the wave function, yielding average expectation values for position, velocity, etc.

Let's consider the single slit diffraction experiment using our model of photon propagation and interaction as depicted in Figure 2. If the photon arrives at the photographic plate at time, $t=0$, it cannot darken the plate. The photon energy and momentum are unavailable for 'real' interaction at this instance. If the photon arrives at time, $t=T/4$, it will slam the photographic plate full force!

The root of the wave particle duality 'problem', lies in the declaration, or decision (i.e. the *interpretation*), that quantum spin is some crazy, special feature of microscopic particles; that it has no counterpart in the macroscopic world; does not, and *could* not, correspond, in *any* possible, conceivable, or imaginable way, to our average, ordinary, everyday, common sense ideas, and notions, of rotation.

In our model, particles are real, spin is real, and particles *really* spin.

Occam's razor to the rescue !

Quantization:

There are no quantum variables that we do not also observe at the macroscopic level. Quantum mechanics deals with mass, spin, charge, magnetic moments, and the conservation of energy and momentum; all phenomena which have a classical counterpart.

In our model, there are no flavors, no colors, no fractional charges, no charged fields, no mass inducing fields (no fields at all!), no ghostly evolution of particle states, no vacuum, no "spin" that doesn't really spin, etc., etc.

As Werner Heisenberg has pointed out, if there were quantum variables that did not correspond to a familiar macroscopic quantity, we would not be able to recognize them.

All the classical concepts and quantities that we use to describe the macroscopic world are quantized at the microscopic level, including mass, the electromotive force, and the magnetic moment of a particle. Furthermore, even acceleration seems to be quantized as evidenced by the electronic transitions in the hydrogen atom, for example.

The only difference between macroscopic, classical physics, and microscopic, quantum physics, is the nature of, and satisfaction of, the boundary conditions of the physical system.

We are allowed to speak of the 'quantization of space'; e.g. in characterizing the electron orbits of the hydrogen atom, but it should be clear this quantization is a property of the relation between the particles, and not a feature of space itself.

Quantized space is *not* chunky.

Ultimately, quantization boils down to the satisfaction of boundary conditions.

Virtual photons can only attach to/interact with an electron when the electron has an effective helicity of zero (during its polarization flip) because virtual photons are spin zero.

This fixes the 'frequency' of the virtual photon coupling two interacting electrons, since virtual interactions, or the exchange of energy and momentum, can only occur when both electrons have helicity $\sim= 0$.

When an electron reaches 'peak helicity', it is able to interact with (i.e. absorb and emit) real photons the most readily and the most efficiently.

The special theory of relativity:

The special theory of relativity is concerned with point-like, spacetime *'events'*, and whether these may be considered simultaneous in various reference frames and in what regard.

In the *universal theory of relativity*, our concern is with the *interaction <u>between</u> particles <u>over some finite amount of time</u>*, and *interactions* are simultaneous in any reference frame.

You cannot do physics with one particle, nor at a single point in space and time.
For every action there is an equal and opposite reaction. In any reference frame.

All interactions occur at the speed of light regardless of the choice of reference frame. *Nature does not know about reference frames.* 'Nature' only 'cares' about the relative positions and velocities of interacting bodies.

The (background of) space in which these interactions occur is fixed, immutable, absolute, and empty; uniform, smooth, flat, and three dimensional; homogeneous, isotropic, and undifferentiated.

Time and space are *absolute*. Only *coordinate systems* are relative.

'Observers' in different inertial reference frames measure the relevant quantities of an interaction using different position *and* time variables; however, they describe the same world, in the same way, with the same results.

The crucial point here is that the different observers measure different *time intervals* **and** different *space intervals* for any given event. This fact takes a bit of the shine off of any mind boggling interpretation of time dilation. If different observers, in different inertial reference frames, measured *different time intervals*, and yet **identical space intervals**, this would be a truly mind boggling break with Galilean relativity.

Most (of my favorite!) authors seem compelled to convince the reader (and themselves?) that time dilation is not merely appearance, but something real, with real, physical, and measurable effects. They all chose to cite the (only?) example of long lived muons resulting from high energy cosmic ray collisions in the earth's upper atmosphere.

In this example, high energy (v/c ~= 0.98) muons are believed to travel longer distances than usual in *our* reference frame because they have a longer lifetime (due to their velocity) in *their* reference frame!

This is clearly nonsense! In the moun reference frame, the muon is at rest, and thus has the usual 'rest mass', or 'proper', lifetime. The fact is the muon is traveling at v/c = 0.98 in *our* reference frame

Time dilation *is* mere appearance, and cannot account for long lived, high energy muons.

In our model, there is no 'phase space' for the decay of high energy muons because they are too massive (14).

The muon must make multiple collisions before reaching 'decay velocity'.

We remind the reader of our theory of muon decay by including the diagram first introduced in "On Parity and Isospin" as Figure 4.

In our model, the muon decays into an electron by 'shedding' excess energy and momentum in the form of its neutrino. The resulting 'propagator' **_must be_** a massless 'virtual lepton'.

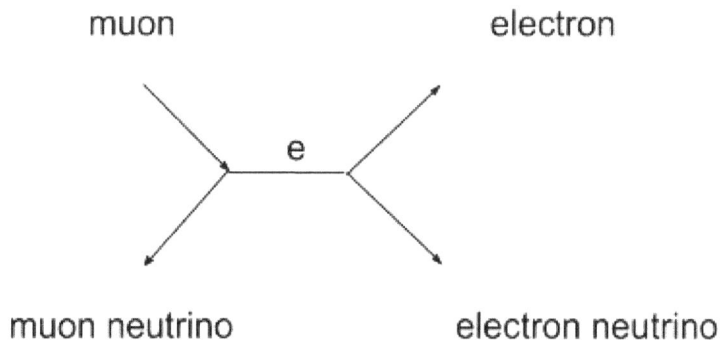

FIGURE 4: Muon decay. The propagator is actually a generic virtual lepton, e/mu/tau.

The energy, momentum, and spin shed by the moun ensures that the 'virtual lepton propagator' has a spin = 0, and is *massless*. This decay process is thus kinematically prohibited for ultra-high energy muons !

The standard model:

In "A Quantum Mechanical Theory of Everything" we said:

The two great pillars of twentieth century physics, general relativity and quantum mechanics, are incompatible and irreconcilable because they are both incorrect.

Of course, we were just having fun with this shopworn cliche.

In our model, general relativity *is* incorrect and we see nothing to recommend it.

On the other hand, we would *not* say quantum mechanics is incorrect except in this particular context. Here, 'quantum mechanics' is taken to mean 'the standard model' and/or quantum field theory, *plus* every possible spooky speculation ever spouted by anyone.

In our view, quantum mechanics is bras, kets, state vectors, wave equations, operators, matrix elements, and expectation values.

Quantum mechanics is *not* wave function collapse, spooky action at a distance, particles appearing two places at once, particles popping out of the vacuum, etc.

In our model, we *reject* quantum field theory while retaining all the rational bits of quantum electrodynamics.

The result is 'simple' relativistic quantum mechanics. The theory of everything.

I believe we have sufficiently proven that gauge theory is incorrect and that gauge symmetry is not a feature of the world (14), that the quantum vacuum is not a thing (6), and that fields of any description are *not* real, but only useful fictions.

Regardless, gauge theory and the magical, miraculous vacuum can no longer come to the rescue of QFT (e.g. renormalization, etc.), so it is a lost cause.

Conclusion:

One *cannot* physically or metaphysically interpret mathematical equations.

The mathematical equations of physics, are *already* (*or should be*) a self-contained and coherent interpretation of all currently known physical data; a codification of the well confirmed, and well understood, observations of how physical systems behave.

Math, as the language of physics, is *already* an interpretation; an expression of knowledge, not easily, or usefully codified, in spoken or written language alone.

The mathematical equations of physics are for making *physical predictions* that can be *tested* by clever and well designed experiments.

They may not be used to make metaphysical assumptions about the world.

Of course, one may fiddle about and accidently discover an equation, then find the equation may be construed to explain several previously known and puzzling facts; the Dirac equation being one of the most famous examples.

The discovery of the Dirac equation was a truly great and momentous invention!

The invention of the Dirac equation was a truly great and momentous discovery!

Truly!

Then, the *interpretations* began, giving us an infinite sea of antiparticles traveling backwards in time (obviously!), just beyond our reach due to some great rift, or potential barrier, between the negative energy quantum vacuum and the positive energy quantum vacuum of our real world.

Better to "make no hypothesis".

the sinister universe:

    iconographic
    apples   !

    (*yawn*)

    no tumbling dice

    no cats,
    ineffably
    entombed

    to die

    (can you even *put* a cat in a box?)

    no black box devices

    no malicious g-d
    no master plan.

    only masses,
    matter

    in motion,
    thru space

    in time

    hand in hand

Books by Greg Feild: The SInister Universe Series

<u>the pentateuch</u>

1. "A quantum mechanical theory of gravitational interactions"
   CreateSpace Independent Publishing, 8/29/2016

2. "Observations on the quantum mechanical nature of gravity"
   CreateSpace Independent Publishing, 10/8/2016

3. "On gravitation and electric charge"
   CreateSpace Independent Publishing, 10/29/2016

4. "On spin, mass, and charge"
   CreateSpace Independent Publishing, 11/29/2016

5. "On angular momentum, acceleration, and absolute motion"
   CreateSpace Independent Publishing, 1/1/2017

<u>the exegeses</u>

6. "The Sinister Universe"
   CreateSpace Independent Publishing, 3/1/2017

7. "On Parity and Isospin"
   CreateSpace Independent Publishing, 4/11/2017

8. "Reflections on the Sinister Universe"
   CreateSpace Independent Publishing, 5/12/2017

<u>the hermeneutics</u>

9. "On Current Physics"
   CreateSpace Independent Publishing, 6/11/2017

10. "A Critical Examination of Classical and Quantum Mechanical Waves"
    CreateSpace Independent Publishing, 6/18/2017

the gospels   :)

11. "On wave particle duality and the quantum of action"
    CreateSpace Independent Publishing, 7/6/2017

12. "On matter, mass, and motion"
    CreateSpace Independent Publishing, 9/14/2017

13. "On action and reaction"
    CreateSpace Independent Publishing, 9/24/2017

14. "A quantum mechanical theory of everything"
    CreateSpace Independent Publishing, 11/5/2017

the compilations

"The Universal Model of Our Sinister Universe: The First Ten Books"
CreateSpace Independent Publishing, 7/2/2017

"The Canons of the Sinister Universe:
The Last Four Books on the Universal Model of Our World"
CreateSpace Independent Publishing, 11/5/2017

Alice's Evidence:

"No, no!", said the Queen.
"Sentence first - verdict afterwards."

"Stuff and nonsense!", said Alice loudly.
"The idea of having the sentence first!"

↺   ↻   ↺   ↻

"Who cares for you?", said Alice.
"You're nothing but a pack of cards!"

-- Lewis Carroll
Alice's Adventures in Wonderland

White Rabbit:

When logic and proportion
Have fallen sloppy dead
And the White Knight is talking backwards
And the Red Queen's off with her head ...

Remember - - - - what the Dormouse said

Feed your head

-- Grace Slick

there is nothing but atoms and the void

On Rotation

Greg Feild

August 18, 2018

On the improvement of the understanding:

> . . . ideas which are clear and distinct can never be false : for ideas of
> things clearly and distinctly conceived are either very simple themselves,
> or are compounded from very simple ideas --- that is, deduced therefrom.
> The impossibility of a very simple idea being false is evident to anyone
> who understands the nature of truth or understanding and of falsehood.

⇔

> . . . we may never, while we are concerned with inquiries into actual things,
> draw any conclusions from abstractions; we shall be extremely careful not
> to confound that which is only in the understanding with that which is in
> the thing itself.

⇔

> . . . words are formed according to popular fancy and intelligence,
> and are, therefore, signs of things as existing in the imagination,
> not as existing in the understanding.

-- Benedict de Spinoza

⇐ ⇒ ⇐ ⇒

The discovery of truth is prevented most effectively, not by the false appearance
things present and which mislead into error, nor directly by weakness of the reasoning
powers, but by preconceived opinion, by prejudice, which as a pseudo *a priori* stands
in the path of truth and is then like a contrary wind driving a ship away from land, so that
sail and rudder labour in vain.

-- Arthur Schopenhauer

Abstract:

In this paper, we show all phenomena can be understood as the conservation of mass, angular momentum and rotational kinetic energy.

In addition, we derive the equation for the relativistic mass of a particle, using only Newton's second law and and the kinematic variables defined by the particle wave function, without resort to, or reference to, coordinate transformations.

We also provide a small correction to our model of photon propagation, and continue our critical examination of the concept of potential energy.

Finally, we demonstrate the equivalence of Newtonian and Hamiltonian dynamics.

In the universal model, nature acts to minimize the work done during any interaction.

Die Welt ist alles, was der Fall ist.

-- Ludwig Wittgenstein

Jabberwocky:

'Twas brillig, and the slithy toves
Did gyre and gimble in the wabe:
All mimsy were the borogoves,
And the mome raths outgrabe.

Humpty Dumpty:

"When I use a word," Humpty Dumpty said, in rather a scornful tone,
"it means just what I choose it to mean -- nothing more nor less."

"The question is," said Alice, "whether you *can* make words mean so many things."

"The question is," said Humpty Dumpty, "which is to be master -- that's all."   :)

-- Lewis Carroll
Through the Looking-Glass

One should not wrongly *materialize* "cause and effect," as the natural philosophers do
(and whoever like them naturalize in thinking at present), according to the prevailing
mechanical doltishness which makes the cause press and push until it "effects" its end:
one should use "cause" and "effect" only as pure *conceptions*, that is to say,
as conventional fictions for the purpose of designation and mutual understanding, --
not for explanation.

-- Friedrich Nietzsche
1886

Neologisms:

We've coined a lot of phrases during the development of "the universal model", mostly on the fly and without too much forethought:

- neutrinium
- the universal model
- the total coupling charge; tcc
- quantum mechanical electromagnetic induction: qemf
- the tetrahedron (perhaps the polyhedron, polyhadron?)
- the electromagnetic charge
- the mass charge
- the compton radius of the neutrino
- mass isospin
- spinoring
- corpusculating

"The universal model" is pretty dull (and already taken by someone!) but it is a handy adjective; e.g the universal reference frame, the universal principle of equivalence, etc.

What we christened mass isospin should really be called charge isospin, and weak isospin (which we had retained for historical reasons) should really be called mass isospin, if we are are to be true and parallel to the original isospin model for nucleons. We will make and justify these changes later in this book.

The terms we chose for the methods of propagation for the photon and electron, corpusculating and spinoring, are rather long, boring, and clumsy.

What we *meant* to say, was, photons gyre and electrons gimble.   :)

But, take your pick ...

Inventing words is hard!

Errata:

I am not an accomplished speller, although I do know the difference between principle and principal; at least in principle! Unfortunately, over our last several books, we have written the phrase 'principle moment of inertia', too many times to count! I blame the spell checker, which is now, of course, screaming bloody murder. But, it's dun! Moving on ...

We will resurrect the ideas of 'the electromagnetic charge' and 'the mass charge' first introduced in "On Parity and Isospin" (along with the idea of 'mass isospin') and later abandoned for technical reasons. These ideas were not working out as we had the concepts of mass and charge isospin exactly backwards, along with the identity of the corresponding conserved currents. Also, of course, our model of the particle magnetic moment was not fully realized at that time.

We also offer a small, but important, clarification on the nature of the photon and how it rolls.

Finally, in all our expressions and formulations we will bring all suppressed variables (i.e. $c=1$, $h^{bar}=1$) to the fore. To date, we've been carelessly mixing Units, Natural Units, etc.; obviously not without some unnecessary confusion!

Introduction:

Relativistic quantum mechanics may be considered the theory of everything.

Rotation, in all its guises; as translation, angular momentum, spin, helicity, and precession, is the basis for all particles and all particle interactions.

Angular momentum is *the* fundamental quantity, and quality of matter, and the only constant of nature.

Angular momentum is *more* fundamental than even mass or electric charge, since these quantities are essentially emergent properties of particle spin.

The fundamental unit of angular momentum is Planck's quantum of action; h.

Everything reduces to spin. Spin is absolute motion. Spin is contagious.

Spin drives the universe. Spin is in.

The most fundamental, irreducible quantity we can assign to an elementary particle is spin. A particle is essentially spin at a point in space and time.

One cannot deny spin!

↺ ↻

gf

:)

Newton's laws:

Our previous reflections on Newton's laws and the Lorentz force are provided in Appendix A for ease of reference. [Appendix removed. Please see "Everything".]

In Appendix A, we generalized, or univeralized, Newton's third law from $\mathbf{F_1} = -\mathbf{F_2}$ to $(action)_1 = -(action)_2$, or mathematically speaking;

$$\mathbf{F_1} \times d_1 \times t = -\mathbf{F_2} \times d_2 \times t \tag{1}$$

Equation (1) states that the *work done* by the particles, each on the other, Is equal and opposite, rather than the forces. The forces are *only* equal in magnitude, if the two objects each move the *same distance* (i.e. $m_1 = m_2$), or for static forces.

So, for example, as depicted in Figure A1, the central force between two bodies in the universal reference frame (i.e. the force on the reduced mass) is not the same as the individual forces of each body on the other, but their sum.

We must conclude Newton's third law ($\mathbf{F_1} = -\mathbf{F_2}$) is incorrect, even for most *central* forces.

I know ! Right ? . . . we, too, are a little shocked, surprised, and unsettled . . .

However, the actions of the two particles are equal and opposite. Action is always conserved.

In order to fully understand the nature of two body motion and the forces involved, we will recast Newton's three laws of motion (of Appendix A!) solely in terms of angular momentum. We believe angular momentum to be *the* fundamental concept or quantity of physics, even more fundamental than linear momentum, which is only a limiting case (... we always learn it backwards in school!).

Linear momentum is a human-sized, *earthbound* concept, as are the ideas of force, rest, and the notion that objects do not have the propensity to spin. If we did not inhabit human bodies on a flat earth, we should probably believe none of these things!

In addition, and 'as we know', one may always 'transfer linear momentum away'. The conservation of linear momentum attests to the fixity of, and the isotropic, homogeneous, etc. nature of, space and time. There is no more physics there.

Angular momentum is where the action is ! (No pun intended.)
Rotational symmetry is the only symmetry.
Angular momentum is the only conserved quantity and determines all the laws of physics.
This is the premise of the universal model.

We will reexpress Newton's laws in terms of angular momentum, rather than linear momentum, with all calculations done in the universal reference frame of Figure A1.

The advantages of this approach are several. When we analyse interactions in the center of mass of the particles of interest, all external torques (arising from conservative gravitational sources) cancel out; these external forces affect only the linear momentum of the center of mass, which is arbitrary.

When we characterize interactions in terms of torques, and the rate of change of angular momentum of each particle in a system of particles relative to an 'arbitrary' point, the results are applicable, and can be generalized, to *all* reference frames, even to *accelerating* reference frames.

That's the power of spin !   :)

Newton's first law becomes:

If a particle does not experience a change in *angular momentum* relative to *any* arbitrarily chosen point in the universal reference frame, then the particle is considered to be free (i.e. there are no net forces acting on it).

For a two body system, if there is no change in the angular momentum of *either body* comprising the two body system relative to *any* arbitrary point (this 'excludes' the choice of points lying along the unit vector, **r**), then the particles are *not interacting*.

Newton's second law becomes:

A particle that undergoes a change in angular momentum relative to our arbitrarily chosen point (i.e. the origin of our coordinate system) is said to experience a net torque, τ ;

$$\tau = dL/dt = \mathbf{r} \times \mathbf{F} \tag{2}$$

where the force, **F**, is defined by equation A1.

For two body 'central force' motion, the torques experienced by each individual body relative to our chosen reference point of Figure A1, are equal and opposite;

$$\mathbf{r_1} \times \mathbf{F_1} = -\mathbf{r_2} \times \mathbf{F_2} \quad ; \quad |\mathbf{F}| = |\mathbf{F_1} - \mathbf{F_2}| \tag{3}$$

Now that we are expressing the forces between interacting bodies in terms of torque, we can include the 'spin-spin' interaction of our theory in a rigorous way into our universal force equation. Previously, we had to *resort to* the concept of potential energy to include this interaction.   :(

We remind the reader that the "Lorentz torque" is defined as the interaction of the moment of inertia of one body with the "gravitational magnetic vector", **B_g**, of the other body, and vice versa.

$$\tau_{SPIN} = I_1 \times B_2 + I_2 \times B_1 \tag{4}$$

and

$$\tau_{TOTAL} = \tau_{FORCE} + \tau_{SPIN} \tag{5}$$

Newton's third law becomes:

During a two body interaction, the two bodies will undergo equal and opposite changes in their respective '*actions*'; i.e. they will have equal, and 'opposite', changes in kinetic energy.

$$\int F_1 \cdot dr_1 dt = -\int F_2 \cdot dr_2 dt \tag{6}$$

Or, since the time is common to both integrals, and the limits of integration are arbitrary;

$$\Delta T_1 = -\Delta T_2 \tag{7}$$

Newton's universal law of gravitation, expressed in the center of mass of a two body system, becomes (equation A9);

$$F/E_{TOT} = K^*(c/R)^2 \mu - K^*(\mu v^2/R^2) - K^*(l^2/\mu R^3) \tag{8}$$

where the second term on the right hand side is the *coriolis* force; our answer to spacetime disturbances.

We note, just for fun, that all test particles follow the exact same trajectory in a Newtonian gravitational potential.

However, the concept of potential energy is *so* last-century.

Potential energy:

The concept of potential energy, and that of the force field and the 'test charge', has always been problematic. Potential energy is defined as the work done to move an object a prescribed distance in a conservative force field. How is this work being done and by whom? More importantly, how is this potential energy recognized, *physically acquired*, stored and released by a particle or a system of interacting particles?

It's not clear.

Potential energy is a useful concept for studying closed, conservative systems, like planetary orbits, the hydrogen atom, etc. However, it is *incorrect* to imagine that some nebulous, ethereal, spooky potential energy is in any way "stored" in these bound systems. The energy of these systems are completely determined by summing the individual kinetic energies of the particles Involved.

For example, in an atomic explosion, one is not releasing stored potential energy. One is freeing kinetic energy. The orbital kinetic energy of the bound particles becomes linear kinetic energy once the particles are released.

The concept of potential energy is particularly ill-defined for understanding how these closed, conservative systems form in the first place. Imagine an electron and a proton separated to a distance, $R = \infty - 1$. Is the potential energy between the two particles zero or infinite? Conventionally, we would say zero, however, when we 'let the two particles go', they will start moving toward each other, acquiring ever increasing, equal and 'opposite' kinetic energy on the way.

In the formation of the hydrogen atom, -13.6 eV of work is not done in bringing an electron in, to orbit around a waiting, stationary proton; rather an electron and proton with the correct relative kinetic energies (and momenta, for direction!) will meet and form a bound system. We would not even venture to say that the proton "captures" a passing electron (except as a simplifying approximation for models, etc.), as this would imply a fixed center of force, or "cause", which we *do not allow* in our model of action and reaction.

Particles act on each other, equally, always.

In order to remove to concept of potential energy from our model, but keep the concept of the conservation of energy, we must allow for 'negative' kinetic energy. The kinetic energy would take the sign of the particle momentum.

Kinetic energy is thought to always be positive because it depends on the square of the velocity, however, this is *only* a convention.

Work can be negative. Potential energy can be negative. Now kinetic energy is negative!

Getting rid of potential energy, formally, at least, has several important consequences for putting the photon on equal footing with the leptons. Photons have kinetic energy, but they cannot have potential energy by the usual definition of bringing a particle in from infinity and calculating the work. However, the photon is allowed to execute closed orbits in a potential well, although the photon has to be 'going in the right direction' before capture. Even in an 'infinite potential', a photon cannot reverse linear direction, but rather the photon frequency will tend to zero.

Hamiltonian dynamics:

In our model, the Lagrangian for *any* interaction, classical or quantum mechanical, is the change in kinetic energy of the system;

$$L = \Delta T = \Delta E = \Delta m c^2 \qquad (9)$$

With this definition, when we move to relativistic quantum mechanics, we will no longer need to "kludge" an effective Lagrangian from guesswork, knowing the desired results in advance.

So, our minimization principle is; nature always minimizes the change in kinetic energy integrated over the time of an interaction,

$$\delta \int \Delta T \, dt = 0 \qquad (10)$$

Or, *equivalently*; nature minimizes the work done over the time of the interaction,

$$\delta \int \int_r \mathbf{F} \cdot d\mathbf{r} \, dt = 0 \qquad (11)$$

Besides the obvious lack of an interaction term (!), equation (10) involves scalar quantities, and we are no longer confident about unambiguously assigning such variables a positive or negative value. So, equation (11) is our formal minimization principle.

(We expect this will lead to a minimization principle of least time for light, and least distance for matter, although we've yet to work through the math.)

Since the limits of the time integral in equation (11) are arbitrary, this means nature is *always* minimizing the work, and it would be nice to have an instantaneous (i.e. derivative) expression for our minimization principle.

We want to rid our theory of all things linear (except in the mathematics!), so we will recast equation (11) in terms of the variables of the universal reference frame of Figure A1, except that we will (explicitly) remove the time integral;

$$\delta \int \tau \cdot d\theta = 0 \tag{12}$$

where the torque is

$$\tau = dL/dt = \mathbf{r} \times \mathbf{F} \tag{13}$$

and if we can write

$$d\theta = (d\theta/dt)\, dt = \omega\, dt \tag{14}$$

then equation (12) becomes

$$\delta \int dL/dt \cdot \omega\, dt = 0 \tag{15}$$

In equation (15), of course, the integral must be done for object 1 and 2.

Now we have a formulation of Hamilton's principle derived from Newton's laws and identical to that resulting from our new universal Lagrangian, *and* cast in terms of *angular frequency*.

Spoiler alert: Foreshadowing!

The conservation of mass:

The first conservation law was the conservation of mass. Next, was added the conservation of energy. Then, these two concepts were merged to form the conservation of mass-energy.

Now we are back to conservation of mass.

Mass is not converted into energy. It is converted into other mass; either light or matter.

Technically, there is no energy, *only* mass and motion.

Mass is a measure of 'particle angular momentum'.

Consequently, we will always speak of the mass of a photon and 'never' the energy.

There are two kinds of mass; light and matter, or photons and leptons.

All quantum mechanical probability currents express the conservation of mass.

Conservation of mass holds in any reference frame, because mass difference is a scalar and Lorentz invariant.

There is only conservation of mass. We retain kinetic energy for practical purposes.

The fundamental atom:

The fundamental atom, the origin and source of all mass and interaction, is Planck's quantum of action; h.

The photon *is* one quantum of action. However, the photon is *not* the most 'fundamental' particle! The sole source of photons is the acceleration of leptons. Thus, the photon is the quantum of *interaction*.

The sole source of all photons is accelerated matter shedding mass, so to understand the photon and the quantum of action, we really need to understand the nature of the leptons; in particular the spin nature of the neutrino and how it 'acquires' mass. Someday!

But, we start with the photon.

The photon:

The photon is one unit of angular momentum, *the fundamental unit of angular momentum*, Planck's quantum of action; h. The angular momentum of the photon is always h, and the speed is always c. So, how does the photon gain and lose energy and linear momentum during an interaction?

In our model, the projection of the photon angular momentum vector, along the direction of travel, varies sinusoidally at the frequency that defines the energy and momentum of the photon according to the usual relations;

$$E = h\nu$$
$$p = h\nu/c \qquad (16)$$
$$m = h\nu/c^2$$

Let us say the photons 'gyres' as illustrated in Figure 1.

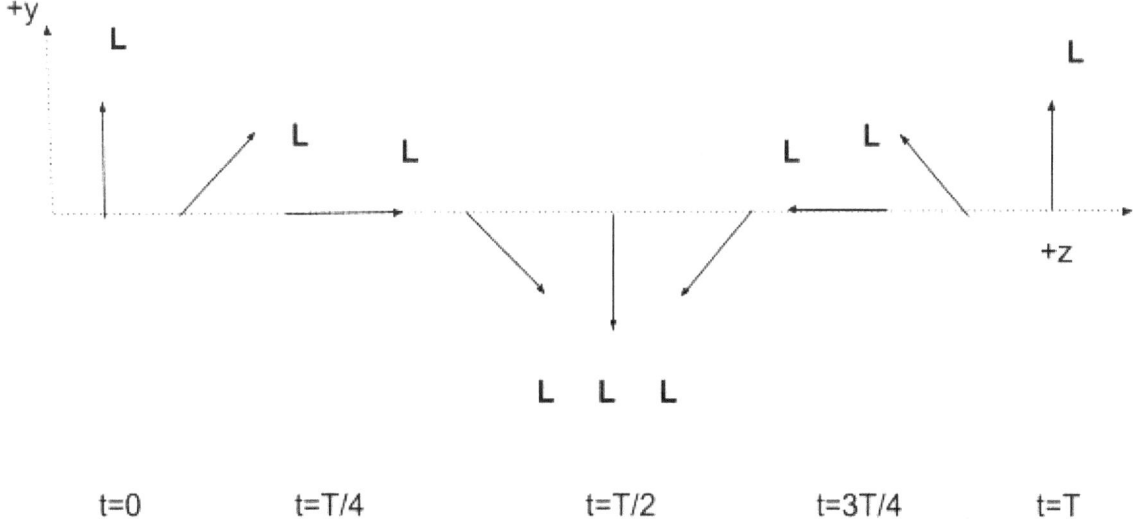

Figure 1: The photon rolls, or 'gyres', along the direction of travel projecting the spin angular momentum vector sinusoidally along the direction of propagation while maintaining a constant polarization and angular momentum; L = h. We could also say that the photon '**spirals**'. At t=0, the plane of the photon spin is *coplanar* with the z-axis.

The photon is massive, yet *inertialess*, because it only has one component of angular momentum, and the projection is along, or parallel to, the direction of propagation.

The force on or due to a photon is

$$F = dp/dt = (h/c)\, d\nu/dt \tag{17}$$

The impulse transferred to or from the photon is

$$P = \int_t (h/c)(d\nu/dt)\, dt = (h/c)(\nu - \nu') \tag{18}$$

We may also construct a minimization principle for the photon;

$$\delta\int W\, dt = \delta\int h\,(d\nu/dt)\, dt = \delta\int \hbar\,(d\omega/dt)\, dt = 0 \tag{19}$$

Just like our new 'classical' minimization principle of equation (15), equation (19) involves only angular variables, and the integration is explicit only for the time of the interaction.

Of, course, to apply these equations to physical situations, one must know the forces in order to minimize the work.

The photon has an antiparticle. We will call it the d'oh-ton!
Photons spin to the left and anti-photons spin to the right.

When a photon and a d'oh-ton collide, they form a virtual photon. The photon has a gravitational charge equal to its mass. In our model photons may interact with one another as well as with the leptons.

We can see from Figure 1, that the amount of angular momentum *available for interaction* (and/or transfer), as well as the *ability of the photon to interact at all*, varies sinusoidally in time as the photon travels through space, hence the appearance and behavior of electromagnetic waves.

We will interpret the scheme portrayed in Figure 1 *literally*, and proclaim the photon is a two-dimensional, planar, flat, circular, spinning disk of fundamental angular momentum, h.

The leptons would then be three-dimensional spinning *blobs* of fundamental angular momentum, with three vector components, only one of which may be projected along the direction of travel; hence the leptons acquire *inertia*.

The neutrino:

We imagine the neutrino to be one quantum of action per unit volume of space, resulting in an available, or 'functional', angular momentum of $h^{bar}/2$.

The neutrino is a spinning point particle with three angular momentum vectors.

Only one component of the angular momentum may be projected along the direction of travel, and the total angular momentum 'gimbles' about this direction, as do the other two components.

The gimbling of the neutrino is illustrated in Figure 2.

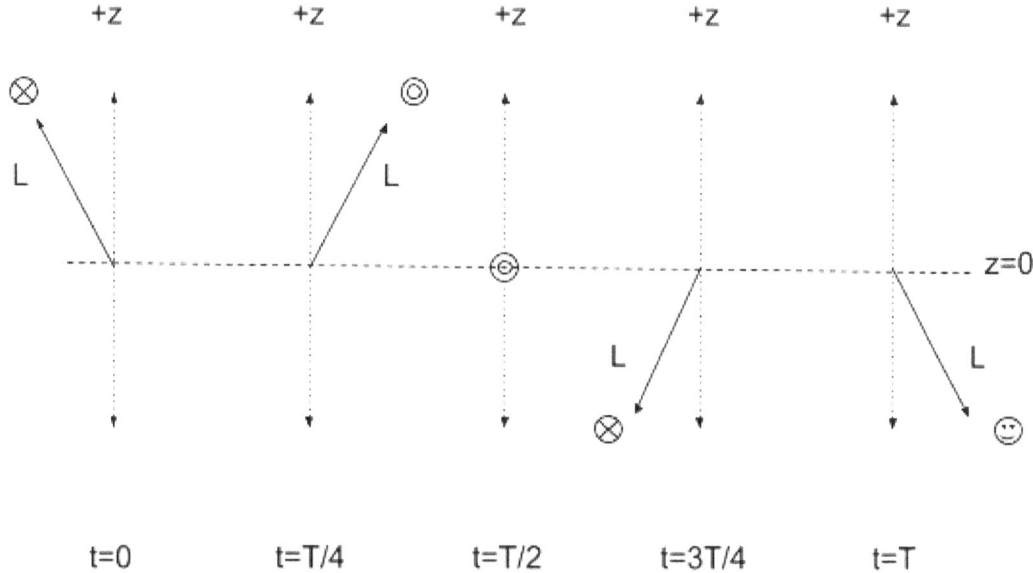

Figure 2: At rest with a lepton traveling the the z-direction. The spin angular momentum vector 'precesses' about the direction of motion, tracing out a closed, three dimensional figure eight. The x symbol represents motion into the page. The dot symbol represents motion out of the page. At time T/2, we see the the angular momentum is *perpendicular* to the direction of travel. (This is when the lepton engages in 'virtual' interactions.)

To represent an antilepton, simply swap the x symbols and the dot symbols.

From Figure 2, we can see that leptons *always* spin to the left, regardless of whether the spin is pointing up or down. That is the power of gimbling!

The neutrino has a gravitational magnetic moment given by (14);

$$\mu = (\hbar/2)(1 + \tfrac{1}{2} v^2/c^2 + \tfrac{3}{8} v^4/c^4 + ...) \tag{20}$$

The magnetic moment is dependent on the particle velocity.

The neutrino magnetic moment may be considered to be a functional 'negative gravitational charge'.
Two neutrinos will attract one another due their mass or gravitational charge.
However, they cannot collide, due to the repulsion arising from their identical spins

Identical particles repel one another, and cannot collide, because they spin in the same direction. Particle antiparticle pairs attract one another, because they *can* collide, merging to form new particles. ;)

The electron:

In our model, the existence of the electron is due to the fundamental and irreducible phenomenon of electromagnetic induction (14). In this model, spinning mass gives rise to the electric charge and the electron magnetic moment.

The mass of the electron is the mass of the neutrino multiplied by the *magnitude* of the electric charge;

$$m_e = e\, m_\upsilon \tag{21}$$

The Coulomb is a 'funny' unit, since it does not involve M, L, or T ! We should like to get rid of it eventually, no offence to Charles de - !, but units of mass, or even no units, would be a better choice. We would then speak of electrical current as mass per second, or even just 'per second' (Hertz)!

The magnetic moment of the electron is velocity dependent like that of the neutrino and is given by;

$$\mu = (e\hbar/2m_e)(1 + \tfrac{1}{2} v^2/c^2 + \tfrac{3}{8} v^4/c^4 + ...) \tag{22}$$

which is just the neutrino magnetic moment multiplied by e/m_e.

The gimbling of the electron is described by the usual wave equation variables;

$$p = h/\lambda \quad ; \quad E = \upsilon h \tag{23}$$

Now, $\lambda = v/\upsilon$, so

$$p = h\upsilon/v \tag{24}$$

Then

$$\frac{dp}{dt} = \frac{\partial p}{\partial v}\frac{dv}{dt} + \frac{\partial p}{\partial \upsilon}\frac{d\upsilon}{dt} \tag{25}$$

$$\frac{dp}{dt} = -\frac{h\upsilon}{v^2}\frac{dv}{dt} + \frac{h}{v}\frac{d\upsilon}{dt} \tag{26}$$

We also know

$$E = h\upsilon = mc^2 \tag{27}$$

and,

$$dm/dt = (h/c^2)\, d\upsilon/dt \tag{28}$$

so equation (26) becomes

$$dp/dt = -(E/v^2)dv/dt + (c^2/v)dm/dt \tag{29}$$

We *also* know

$$p = mv \tag{30}$$

and

$$dp/dt = v\, dm/dt + m\, dv/vt \tag{31}$$

Equating equations (29) and (31) we have

$$v\, dm/dt + m\, dv/vt = -(E/v^2)dv/dt + (c^2/v)dm/dt \tag{32}$$

We do a boatload of algebra and obtain

$$dm/m = (1 + v^2/c^2)/(1 - v^2/c^2)\, dv/v \qquad (33)$$

Then we realize we are unable to do the integral ... :(

A homework assignment !

(That's all you get for ten dollars!) :)

    Math is hard.

It takes a universe ...

Duffman says a lot of things, oh yeah!

-- Duffman
The Simpsons

:)

The fundamental lepton:

We have imagined the electron, muon, and tau, to be three manifestations of one fundamental lepton; the tetrahedron. In our model, the tetrahedron has three principal moments of rotational inertia $I_{e,mu,tau}$, corresponding to the three lepton masses.

The gimbling of a particle's moment of inertia, I, about the direction of propagation (in concert with the angular momentum vector, $L = \sqrt{3}\, \hbar/2$, *is* relativistic mass.

Curiously, this moment of inertia seems to *not* contribute to particle angular momentum, but *only* to the mass. Let's explore! What could this moment of inertia be?

Consider a non-relativistic particle with velocity v, and kinetic energy

$$T = \tfrac{1}{2} m v^2 \qquad (34)$$

$$T = \tfrac{1}{2} m \lambda^2 \nu^2 \qquad (35)$$

Now, if we define the moment of inertia I to be

$$I = m \lambda^2 / (2\pi)^2 \qquad (36)$$

then we can write

$$T = \tfrac{1}{2} I \omega^2 \qquad (37)$$

We define the 'radius' of a point particle to be equivalent to the wavelength. This radius travels 2π radians every wave cycle. The moment of inertia is dependent on the particle wavelength, $I = I(\lambda)$.

Is there a corresponding angular momentum, $L = I\omega$? It seems there should be! In our model, the magnetic moment increases with velocity and this must be the source. So, we define the angular momentum of an elementary particle to be

$$L = \sqrt{3}\, \hbar/2 + I\omega \qquad (38)$$

and

$$E = m_0 c^2 + \tfrac{1}{2} I \omega^2 \qquad (39)$$

We can also solve for the 'rest frequency' of an electron;

$$m_e c^2 = \hbar \omega_0 \tag{40}$$

$$\omega_0 = m_e c^2 / \hbar = (2\pi c)/(h/m_e c) \tag{41}$$

$$\omega_0 = 2\pi c / \lambda_e \tag{42}$$

where λ_e is, of course, the Compton wavelength of the electron, which we also take to be the 'rest radius'. The uncertainty principle can provide us with the 'rest momentum',

$$\lambda_e \, \Delta p_e \approx \hbar \tag{43}$$

We can understand these uncertainties, since, even an electron 'at rest', is gimbling, with the radius varying sinusoidally, and the momentum apparently flipping direction at the same rate.

In a similar manner, we can assign a functional radius to the photon as well.

Now, we can offer a **bold**, new hypothesis concerning the double slit experiment.

Perhaps, somehow, when the radius of a particle is on the order of the slit separation, the particle will pass through one slit, whilst 'clipping' or 'grazing' the other slit, altering its angular momentum. Somehow ...

But, back to the tetrahedron, which may be merely metaphorical !

We can use the SU(3) rotation matrices of the standard model to operate on our inertia tensor, and translate from one lepton flavor to another. We will refer to these as rotations in *charge isospin* space.

In addition, we imagine that rotations in 'weak isospin' or SU(2) space, transform the charged lepton into its corresponding neutrino.

Since our model does not involve the weak charge, we must reluctantly rename the concept of weak isospin. We shall now call weak isospin, *mass isospin*.

These ideas are summarized in Table 1.

The leptonic table:

LEPTONS   ANTI-LEPTONS

electron	electron neutrino	PARITY ⇔	electron antineutrino	positron
⇐	CHARGE	MASS ⇕	CHARGE	⇒
muon	muon neutrino	PARITY ⇔	muon antineutrino	anti-muon
⇐	CHARGE	MASS ⇕	CHARGE	⇒
tau	tau neutrino	PARITY ⇔	tau antineutrino	anti-tau
⇔	mass isospin	charge isospin ⇕	mass isospin	⇔

TABLE 1: The leptons and their interrelations; or the kleptogenesis of the leptoquarks.

Any lepton can be 'generated' from any other by the appropriate applications of the parity operator, the mass isospin operator, and our newly proposed 'charge isospin' operator.

The electron, muon, and tau have the same electric charge, but different masses.

Each lepton and its neutrino has the 'same mass', except for an additional (multiplicative) electric charge.

The parity operator transforms particles into antiparticles.

That is the premise of the universal model.

That is the power of symmetry !   :)

Now, let's give the photon a moment of inertia and a magnetic moment. We can see from equations (20 and (22) that a particle's magnetic moment is the angular momentum times a multiplicative constant. The photon angular momentum is h, so the magnetic moment is

$$\mathbf{m}_\gamma = h\,\mathbf{i} \quad ; \quad \mathbf{i} = \mathbf{v}/|\mathbf{v}| \quad ; \quad v = c \tag{44}$$

The mass of the photon is given by equation (16); we insert this into equation (36) to obtain the moment of inertia of the photon

$$I_\gamma = (h\nu/c^2)(\lambda^2/4\pi^2) \tag{45}$$

The electromagnetic charge:

Interactions conserve spin, mass, and charge. In this section, we will generalize the concepts of electric charge and mass, to include all conserved quantities.

Therefore, we define the electromagnetic charge to be

$$Q_{EM} = e\hbar/2c\,\mathbf{s} \tag{46}$$

and the mass charge to be

$$Q_{MASS} = m\hbar/2c \tag{47}$$

These quantities must be conserved at every 'vertex'.

We are still musing over the potential utility of these definitions …

They are summarized in Table 2.

Force	Coupling constant	Conserved current	Rotation basis	Conserved charge
Electricity	e/m_e	Mass isospin	e, mu, tau	$e\hbar/2c\,\mathbf{s}$
Gravity	m_e/e	Charge isospin	e, ν_e	$m\hbar/2c$

TABLE 2: Table of coupling constants, conserved currents, and charges: Note: $m_e/e = m_\nu$

Muon decay:

In our model, muon decay *involves no photons* (or any other 'gauge' boson) and is a fundamentally different type of decay from beta decay, or other decays involving a *bound system of particles*. The universal model of muon decay is shown in Figure 3.

There is no 'coupling charge' *per se*, at the vertices, except for mass, for the calculation of lifetimes, decay rates, and probabilities of decay. The only requirement is the 'propagator' be massless. This fixes the phase space for the energy and momentum of the three final state particles, given a particular muon energy.

We imagine the tetrahedron is rotating, or dropping, from the muon axis to the electron axis!

In our model, the muon 'sheds' energy and spin in the form of its neutrino. The energy and spin shed ensures that the ensuing virtual lepton propagator has spin = 0, and is *massless*. In our model, all 'propagators' are massless.

This requirement places constraints on the energy and momentum of the initial and final states in muon decays, and should help to explain the energy hierarchy of the three particle families.

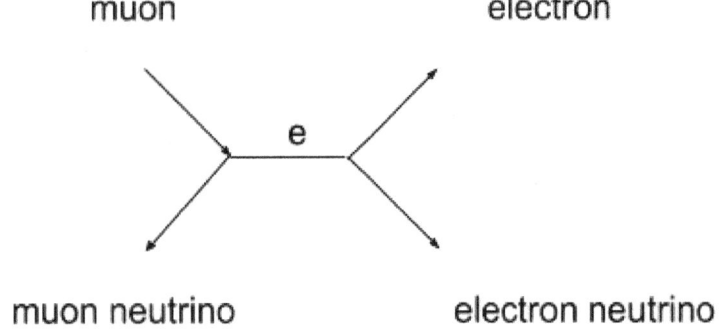

FIGURE 3: Muon decay. The 'propagator' is a generic, 'virtual' lepton, e/mu/tau.

Pilot wave theory:

Our model of electron propagation may be considered a "reverse pilot wave theory". Let us call it the 'pilot particle theory', or perhaps, the 'pilot velocity vector theory'.

In our model, the electron travels a well defined path dictated by the direction of the velocity and/or momentum vector. The velocity *and* momentum of the electron are *completely and solely determined* by the energy, or *frequency*, of the particle.

The electron angular momentum vector gimbles about the 'center of momentum', varying sinusoidally in the ability to engage in 'real' and/or 'virtual' interactions.

Every time the "electron wave function passes through zero", the electron gets updates from all the matter it is interacting with 'at a distance' and it responds accordingly.

This model of electron propagation explains how electrons can 'tunnel' through 'potential barriers". In addition, we can see how an electron with the proper frequency and *phase*, would travel freely through a regular crystalline lattice,

The hydrogen atom:

Let's consider the formation of an hydrogen atom under idealized conditions, as shown in Figure 4. In a two particle universe, we find an electron and a proton *very* far apart. Still, they note a universal attraction; a force that varies with separation, and so they begin moving slowly, inexorably, toward one another, accelerating as they go, gaining kinetic energy due to the mutual forces of one on the other. These forces are equal and opposite, *however* the particle masses are not, so the particles will travel different distances over the allotted interaction time, gaining equal kinetic energy and opposite momenta.

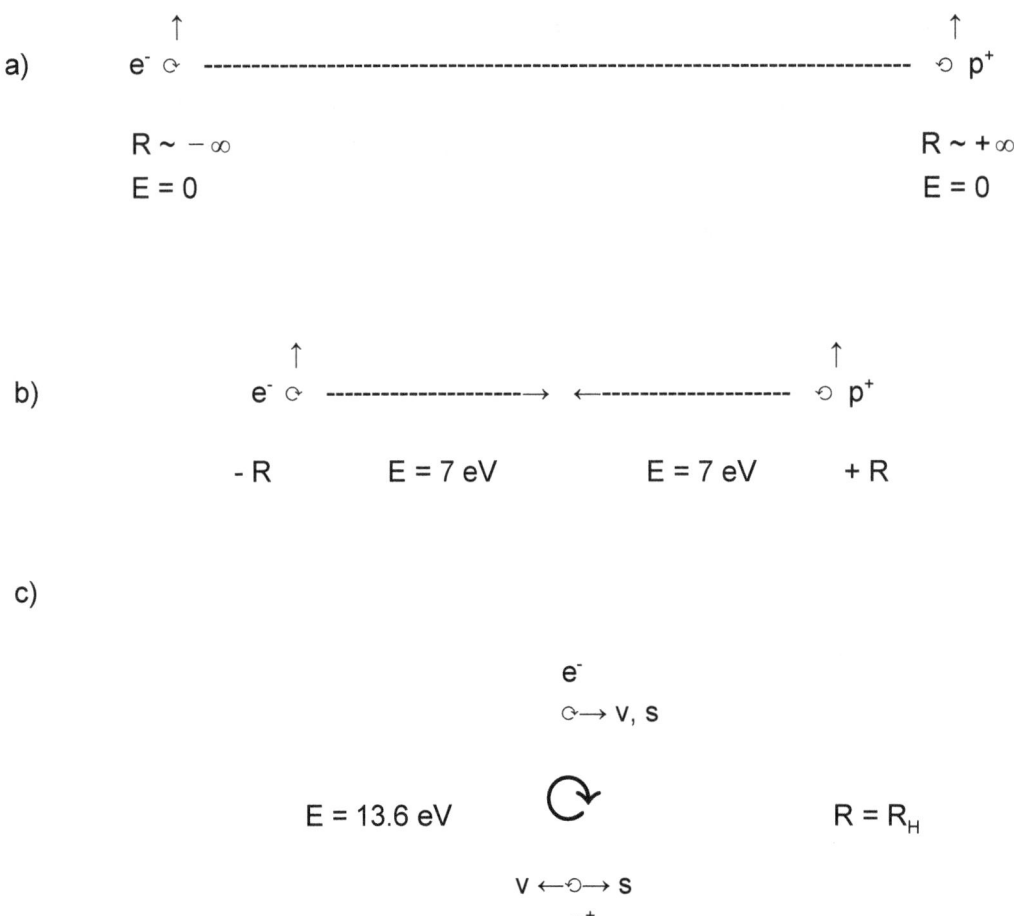

Figure 4: The formation of an hydrogen atom. In a) and b) the spin is shown as pointing up, and the spin *projection* is *not* displayed. In c) we show the projection of the spin and velocity vectors of the particles along, or tangential to, the direction of motion. The electron is matter and the proton is antimatter, so both spins are pointing 'up'!

Given a fortuitous choice of R $\sim= \infty$, we find the electron and proton will have the appropriate energies to form hydrogen when they meet, entering into a mutual orbit about a common center of mass.

The work done on the particles is equal and opposite, resulting in equal and opposite momenta, but a net *gain* in kinetic energy. In the bound system, these kinetic energies correspond to orbital velocities of opposite direction, and they sum to the energy of the hydrogen atom.

We also note the electron and positron have equal and opposite helicities, and these cancel out.

Not only do the 'static' charges attract, but the spins (magnetic moments) of the electron and positron attract too! The two particles cannot be at the same place at the same time, nor can they annihilate, so they orbit.

Perhaps we might think of the vector summation of the work as always summing to zero, and take the scalar sum of the work to represent the total energy of the system.

Now, if we fix our coordinate system at the center of the the proton, the electron orbit will be an ellipse with one focus fixed at the proton center.

In a two particle universe, the hydrogen atom will be *planar*. The electron executes closed elliptical orbits. There are no external forces to change the angular momentum of the particles.

The electron is not a 3D cloud of probability. The proton would be a similar cloud, if so, eh?

Conclusion:

    Potential energy was one the first abstract concepts developed by physicists, and one of the first to assume the mantle of reality; truly believed to be a fundamental aspect of the world; a real, absolutely conserved quantity of energy, somehow taken up and lost by objects during interaction.

    It seems fitting this idea should be the last to fall; this ghostly vestige of *classical* physics.

    Just like our old friend the phlogiston, potential energy turns out to be mass in motion.

    Everything is mass in motion.

    All is action and reaction.

⇐ ⇒ ⇐ ⇒

    Motion is the mode of existence of matter. Never anywhere has there been matter without motion, nor can there be. Motion in cosmic space, mechanical motion of smaller masses on the various celestial bodies, the motion of molecules as heat or as electrical or magnetic currents, chemical combination or disintegration, organic life - at each given moment each individual atom of matter in the world is in one or other of these forms of motion, or in several forms of them at once. All rest, all equilibrium, is only relative, and only has meaning in relation to one or other definite form of motion . . .

Motion without matter is just as unthinkable as matter without motion.

Motion is therefore as uncreatable and indestructible as matter itself;

-- Friedrich Engels
1878

Stimmt!

Everything is connected:

Everything is connected ! :)

Every particle in the universe is constantly communicating the following 'information' to each and every other particle, "my energy and momentum with respect to yours is X."

More specifically, each *particle pair* is sharing such *relative* information.

This communication is neither magical or mystical. There is certainly *no room* in these channels for any more complicated message than relative energy, momentum, and *position*.

These messages are being sent over "dedicated lines"; i.e the virtual photon. There are no intervening fields, and the dedicated lines do not split into particle antiparticle pairs "along the way".

The 'distance' between a particle pair is given by the frequency of the virtual photon; a direct consequence of the $1/R^2$ law. The frequency of the virtual photon determines the *rate* of energy and momentum exchange.

The world is everything that is the case. -- Ludwig Wittgenstein

On math and physics:

Mathematics does not describe the world, mathematics describes our observations of the world.

Nature knows neither numbers nor nullity; is as indifferent to the integer as the infinite, noting nothing but matter in motion.

First Principles:

Descending to a more concrete view [of the law of co-operation], we saw that the law sought must be the law of the continuous re-distribution of Matter and Motion. The changes everywhere going on, from those which are slowly altering the structure of our galaxy down to those which constitute a chemical decomposition, are changes in the relative positions of component parts; and everywhere necessarily imply that along with an new arrangement of Matter there has arisen a new arrangement of Motion.

Hence it follows that there must be a law of the concomitant re-distribution of Matter and Motion which holds of every change, and which, by thus unifying all changes, must be the basis of our Philosophy.

-- Herbert Spencer
1862

. . . perhaps there is nothing about so-called educated people and believers in "modern ideas" that is as nauseous as their lack of modesty and the comfortable ignorance of their eyes and hands with which they touch, lick, and finger everything; and it is possible that even among the common people, among the less educated, especially among peasants, one finds today more *relative* nobility of taste and tactful reverence than among the newspaper-reading *demi-monde* of the spirit, the educated.

-- Friedrich Nietzsche
Beyond Good and Evil

Alles klar!

Observations on the quantum mechanical theory of everything:

1. Everything that is not mandatory, is *expressly* forbidden.

2. Modern physics is neither modern *nor* physics, in the *classical* sense;
   i.e. a rational investigation into the nature of matter and interaction
   based on reason.

3. Just because the wave function is mysterious *to us* (its creators, constructors,
   and inventors!), does *not* mean it represents some mysterious process, it only
   means we have yet to divine the process it does represent.

4. The fabric of spacetime was created out of whole cloth;
   the quantum vacuum from thin air.

5. Today's "modern physics" is an atavism; the four forces for the four
   elements (always four!); spirit guides for motive force.

6. Physics is supposed to be an empirical science, *the* empirical science,
   yet empiricism lay dying in the ditch (with the cart!), if it is not dead already.

7. The modern physicist is a *sadly* deluded idealist. Disillusioned with the apparent
   world, which they cannot understand, they invent fantastic realms; many worlds,
   multiple universes, and manifold dimensions --- these modern mythologists
   and cynical deists!

8. Interactions are deterministic, but this does not mean they are *predetermined*.

9. After 100 years, modern physics is *still* inchoate, incoherent, and growing inexplicably
   ever more complicated and confused, a sprawling morras of mysticism and metaphysics,
   blithely bundled by strapping tape and baling wire, a monstrous Rube Goldberg machine
   of some cosmic (or perhaps, human?) mind.

    10. What's that fudgy smell? --- renormalization.

:(

Resources:

Atomic and Quantum Physics
H. Haken, H.C. Wolf

Modern Elementary Particle Physics
Gordon Kane

Classical Dynamics of Particles and Systems
Jerry B. Marion

Foundations of Electromagnetic Theory
John R. Reitz, Frederick J. Milford, Robert W. Christy

Quantum Physics
Rolf G. Winter

Gauge Theories in Particle Physics
I. J. R. Aitchison and A. J. G. Hey

Quarks and Leptons: An Introductory Course in Modern Particle Physics
Francis Halzen, Alan D. Martin

Quantum Field Theory
F. Mandl, G. Shaw

Theoretical Mechanics of Particles and Continua
Alexander L. Fetter, John Dirk Walecka

and

Elementary Modern Physics						(Best Book Ever!)
Richard T. Weidner, Robert L. Sells

Books by Greg Feild: The SInister Universe Series

the pentateuch

1. "A quantum mechanical theory of gravitational interactions"
   CreateSpace Independent Publishing, 8/29/2016

2. "Observations on the quantum mechanical nature of gravity"
   CreateSpace Independent Publishing, 10/8/2016

3. "On gravitation and electric charge"
   CreateSpace Independent Publishing, 10/29/2016

4. "On spin, mass, and charge"
   CreateSpace Independent Publishing, 11/29/2016

5. "On angular momentum, acceleration, and absolute motion"
   CreateSpace Independent Publishing, 1/1/2017

the exegeses

6. "The Sinister Universe"
   CreateSpace Independent Publishing, 3/1/2017

7. "On Parity and Isospin"
   CreateSpace Independent Publishing, 4/11/2017

8. "Reflections on the Sinister Universe"
   CreateSpace Independent Publishing, 5/12/2017

the hermeneutics

9. "On Current Physics"
   CreateSpace Independent Publishing, 6/11/2017

10. "A Critical Examination of Classical and Quantum Mechanical Waves"
    CreateSpace Independent Publishing, 6/18/2017

the gospels   :)

11. "On wave particle duality and the quantum of action"
    CreateSpace Independent Publishing, 7/6/2017

12. "On matter, mass, and motion"
    CreateSpace Independent Publishing, 9/14/2017

13. "On action and reaction"
    CreateSpace Independent Publishing, 9/24/2017

14. "A quantum mechanical theory of everything"
    CreateSpace Independent Publishing, 11/5/2017

the compilations

"The Universal Model of Our Sinister Universe: The First Ten Books"
CreateSpace Independent Publishing, 7/2/2017

"The Canons of the Sinister Universe:
The Last Four Books on the Universal Model of Our World"
CreateSpace Independent Publishing, 11/5/2017

15. "On Interaction
    CreateSpace Independent Publishing, 4/21/2018

About the author:

I earned a PhD in experimental high energy physics from the Pennsylvania State University working on HERA at DESY in Hamburg, Germany studying photoproduction and deep inelastic scattering in electron-proton collisions.

I did my postdoctoral studies with Yale University working at Fermilab on the CDF experiment at the Tevatron. My primary research interest was particle hadronization in charmonium production in proton-antiproton collisions.

Life, the Universe and Everything:

photons; oscillating! tetrahedrons;
flipping over. orbital decays

physics!

Units and dimensions:

There is essentially only one *variable* of physics; frequency.

Frequency determines the velocity, and hence the linear momentum, mass, and/or kinetic energy of all particles. *Relative* frequencies determines the strength of all particle interactions. We can also characterize macroscopic interactions solely in terms of angular momentum and frequency change.

The fundamental parameters of nature are: t, h, c !

Planck's constant, h, contains *all* the kinematic/dynamical units; M, L, T.

The world is dimensional and our constants should be dimensional too.

For the three coupling constants we choose; G, m_e, e.

The constant ε_0 can be written in terms of G or vice versa. It seems better to cast G in terms of ε_0 (of the vacuum) since ε varies according to the medium of propagation whereas G does not.

As for the electric charge, we suggest and believe it should have the units of mass. This would be an extra "electrical mass" charge that does not contribute to the gravitational mass charge, or its equivalent, the inertial mass.

There are several benefits to this slightly mind sloshing idea. The first being that the formulas for the neutrino magnetic moment, equation (20), and the electron magnetic moment, equation (22), would have the same units; namely, h; a nice symmetry.

Secondly, this would reduce our model to one parameter with units, h, and one variable with units, the particle velocity, v. In addition, the following four quantities would become unitless parameters;

$$v/c, \quad e/m_e, \quad \mu_0, \quad G/\varepsilon_0$$

The Ampere would take on units of mass/second and the 'electrical' current would be

$$I = (e/m_e) \, dm/dt \tag{48}$$

easily allowing treatment of relativistic electrical currents.

Revenge of the Sinister Universe:

The Reality of Everything

Greg Feild

September 5, 2018

About the author:

I earned a PhD in experimental high energy physics from the Pennsylvania State University working on HERA at DESY in Hamburg, Germany studying photoproduction and deep inelastic scattering in electron-proton collisions.

I did my postdoctoral studies with Yale University working at Fermilab on the CDF experiment at the Tevatron. My primary research interest was particle hadronization in charmonium production in proton-antiproton collisions.

Of miracles:

It will be sufficient to observe that our assurance in any argument of this kind [the testimony of men] is derived from no other principle than our observation of the veracity of human testimony, and of the usual conformity of facts to the reports of witnesses.

It being a general maxim, that no objects have any discoverable connexions together, and that all inferences, which we can draw from one to another, are founded merely on our experience of their constant and regular conjunction; it is evident, that we ought not to make an exception to this maxim in favour of human testimony, whose connexion with any event seems, by itself, as little necessary as any other.

Were not the memory tenacious to a certain degree; had not men commonly an inclination to truth and a principle of probity, were they not sensible to shame, when detected in a falsehood: Were not these, I say, discovered by *experience* to be qualities, inherent in human nature, we should never repose the least confidence in human testimony.

-- David Hume
An Enquiry Concerning Human Understanding

Abstract:

In this book, we continue to construct, critique, correct, and compactify our views on "The Sinister Universe"; last visited in the paper "On Rotation".

In particular, we propose two principles to explain all phenomena; the minimization of work, and the conservation of the three, classical and quantum mechanical, 'first integrals of the motion'; E, L, L_z.

perhaps there is only one principle -

action or motion

take your pick!

physics is fun!

Units and dimensions:

Heaviside Units, Heaviside Units (c=1), rationalized Heaviside-Lorentz units (c.g.s), rationalized Gaussian Units, Natural Units, . . .

Confounding and confusing!

In our last book, "On Rotation", we said;

Finally, in all our expressions and formulations we will bring all suppressed variables (i.e. c=1, h^{bar}=1) to the fore. To date, we've been carelessly mixing Units, Natural Units, etc.; obviously not without some unnecessary confusion!

Then, I lost my nerve, as I was still a bit uncertain, given the almost universal tendency of most authors to set c=1, sometimes without explicitly saying so; e.g. Best Book Ever!   :(

Now, we state with confidence, the magnetic moment of the electron (Standard Model) is

$$\mu_e = (e\, h^{bar})/(2\, m_e\, c) \qquad (a)$$

We will also, sometimes, find it useful to break the factor of 2π free from the factor h^{bar}. The origin of the factor 2π is for the conversion of frequency to angular frequency in the particle wave function. However, it also seems to have some significance as a space factor, representing the solid angle integral, or something like that. But, more on this later!

In our new, universal model, the magnetic moment of the electron is (6,13,14,18);

$$\mu = (eh^{bar}/2m_e c)(1 + \tfrac{1}{2}\, v^2/c^2 + \tfrac{3}{8}\, v^4/c^4 + \dots) \qquad (b)$$

and the magnetic moment of the neutrino is

$$\mu = (h^{bar}/2c)(1 + \tfrac{1}{2}\, v^2/c^2 + \tfrac{3}{8}\, v^4/c^4 + \dots) \qquad (c)$$

Similarly, the electromagnetic coupling 'constant', alpha, is (14);

$$\alpha = \alpha_0 (1 + (v/c)^2 + (v/c)^4 + \dots) \qquad (d)$$

$$\alpha_0 = e^2/4\pi\varepsilon\, h^{bar} c \qquad (e)$$

The gravitational coupling constant is

$$\alpha_G = (m_e^2 G)/(\hbar c)(1 + (v/c)^2 + (v/c)^4 + \ldots) \tag{f}$$

The weak coupling constant is

$$\alpha_W = (m_\nu^2 G)/(\hbar c)(1 + (v/c)^2 + (v/c)^4 + \ldots) \tag{g}$$

And finally, the strong coupling constant (14) is

$$\alpha_S = (G/4\pi\varepsilon)^{1/2}(2m_e/\hbar c)(1 + \tfrac{1}{2} v^2/c^2 + \tfrac{3}{8} v^4/c^4 + \ldots) \tag{h}$$

In the universal model, all the 'constants' run, because *the fundamental coupling charge of a particle is the relativistic mass-energy of the particle.*

As particle velocities near the speed of light, the expansion terms in each of the four 'standard model' couplings will begin to dominate, eventually dwarfing the constant, or 'rest mass', terms.

The scale at which the four standard model forces 'merge', or become equivalent, will boil down to a matter of taste, dependent on one's choice of v/c.

In our model, there is no particle 'self-interaction', and no charge screening involved in the interaction between particles.

In our model, there are no fields, no vacuum, and no spacetime.

Never, ever, ever, has a virtual particle-antiparticle pair, ever appeared, anywhere, ever.

Not even 'near' a 'black hole'.

The mind of g-d:

a conservative accountant

The electromagnetic charge:

Interactions conserve spin, mass, and charge. In the universal model, we have generalized the concepts of electric charge and mass, to include all conserved quantities.

We define the electromagnetic charge to be

$$Q_{EM} = e\hbar/2c\ \mathbf{s} \tag{i}$$

and the mass charge to be

$$Q_{MASS} = m\hbar/2c \tag{j}$$

These quantities must be conserved at every 'vertex'.

The nature of these new charges are summarized in Table 1.

Force	Coupling constant	Conserved current	Rotation basis	Conserved charge
Electricity	e/m_e	Mass isospin	e,mu,tau	e*hbar/2*c **s**
Gravity	m_e/e	Charge isospin	e, ν_e	m*hbar/2*c

TABLE 1: Table of coupling constants, conserved currents, and charges: Note: $m_e/e = m_\nu$

Introduction:

Our construction of the universal model is finally coming to a close in this seventeenth (!) book.

All physical phenomena can be accounted for, explained by, or represented as, the minimization of work, and the conservation of energy and angular momentum.

This formulation accounts for all classical and quantum interactions.

This model also explains the nature and behavior of the elementary particles, and the energy hierarchy of the three particle families.

let's do physics !

:)

Newton's laws:

In this section we present a synopsis of the 'universal' formulation, or expression, of Newton's laws, as proposed in our last book, "On Rotation".

Newton's first law:

If a particle does not experience a change in *angular momentum* relative to *any* arbitrarily chosen point in the universal reference frame, then the particle is considered to be free (i.e. there are no net forces acting on it).

For a two body system, if there is no change in the angular momentum of *either body* comprising the two body system relative to *any* arbitrary point (this 'excludes' the choice of points lying along the unit vector, **r**), then the particles are *not interacting*.

Newton's second law:

A particle that undergoes a change in angular momentum relative to our arbitrarily chosen point (i.e. the origin of our coordinate system) is said to experience a net torque, τ ;

$$\tau = d\mathbf{L}/dt = \mathbf{r} \times \mathbf{F} \tag{1}$$

where the force, **F**, is defined by

$$\mathbf{F} = d\mathbf{p}/dt = d(m\mathbf{v})/dt = m\, d\mathbf{v}/dt + \mathbf{v}\, dm/dt \tag{2}$$

For two body 'central force' motion, the torques experienced by each individual body relative to our chosen reference point, are equal and opposite;

$$\mathbf{r}_1 \times \mathbf{F}_1 = -\mathbf{r}_2 \times \mathbf{F}_2 \quad ; \quad |\mathbf{F}| = |\mathbf{F}_1 - \mathbf{F}_2| \tag{3}$$

Newton's third law:

During a two body interaction, the two bodies will undergo equal and opposite changes in their respective '*actions*'; i.e. they will have equal, and 'opposite', changes in kinetic energy.

$$\delta \int d\mathbf{L}/dt \cdot \omega \, dt = 0 \qquad (4)$$

where

$$\tau_{TOTAL} = \tau_{FORCE} + \tau_{SPIN} \qquad (5)$$

and

$$\tau_{SPIN} = \mathbf{l}_1 \times \mathbf{B}_2 + \mathbf{l}_2 \times \mathbf{B}_1 \qquad (6)$$

Newton's universal law of gravitation, expressed in the center of mass of a two body system, becomes;

$$\mathbf{F}/E_{TOT} = K*(c/R)^2 \mu - K*(\mu v^2/R^2) - K*(l^2/\mu R^3) \qquad (7)$$

$$K = G/c^2 \qquad (8)$$

where the second term on the right hand side of equation (7) is the *coriolis* force; our answer to spacetime disturbances.

Since our new coriolis force term goes as $1/R^2$, the classical and quantum mechanical conservation of the the first three integrals of the motion, E, L, L_z, is still guaranteed.

Conservation of energy and momentum:

The two algebraic relations of relativistic physics

$$E = T + m_0c^2 \quad (9)$$

$$E^2 = (pc)^2 + (m_0c^2)^2 \quad (10)$$

can only be realized using "complex algebra". In the Lorentz transformation derivation, the imaginary part arises or is due to a 'rotation of z about the axis ict'. In the Dirac equation derivation, we must employ the Pauli matrices which introduces the factor i.

The simultaneous satisfaction of equations (9) and (10) can also be achieved if we define the total energy to be a complex number

$$E = m_0c^2 + ipc \quad (11)$$

This should not be too shocking in this day and age as almost everything is imaginary or complex (e.g. wave functions).

The factor pc is the kinetic energy, as we shall see, and reduces to $p^2/2m$ in the nonrelativistic limit. The factor of i arises in a 'natural' way when we recast equation (11) as an operator

$$E_{OP} = (m_0c^2 - ic\hbar \cdot \nabla) \quad (12)$$

$$E^2_{OP} = (m_0c^2)^2 + c^2\hbar^2 \cdot \nabla^2 \quad (13)$$

We can now define the kinetic energy operator

$$T_{OP} = -ic\hbar \cdot \nabla \cdot \quad (14)$$

Elementary particles:

In our last book, "On Rotation", we demonstrated the energy and momentum of an electron can be written

$$L = \sqrt{3}\hbar/2 + I\omega \qquad (15)$$

$$E = \hbar\omega_0 + I\omega^2 \qquad (16)$$

$$\omega_0 = 2\pi c/\lambda_0 \qquad (17)$$

where λ_0 is the Compton wavelength of the electron, which we also take to be the 'rest radius'. There is no longer a factor of ½ in the equation for the energy (16) as these are the relativistic expressions. We also need the following relations

$$\hbar\omega_0 = m_0 c^2 = I_0 \omega_0^2 \qquad (18)$$

$$I_0 \omega_0 = \sqrt{3}/2\, \hbar \qquad (19)$$

We define the moment of inertia $I(\lambda)$ to be

$$I = m\lambda^2/(2\pi)^2 \qquad (20)$$

The same relationships hold for the muon and the tau.

Now, here comes the exciting bit. For particles at rest

$$I_e \omega_e = I_\mu \omega_\mu = I_\tau \omega_\tau = \sqrt{3}/2\, \hbar \qquad (21)$$

We see that I has units of $\hbar_{bar}$ or $L \cdot t$.

An elementary particle of constant velocity conserves the three first integrals of the motion; E, L, L_z. The component of angular momentum L_z is projected along (i.e. parallel with) the velocity vector or the direction of travel. The total angular momentum, L, precesses about the direction of travel at a constant frequency.

Particle families:

In the universal model, the processes of elementary particle production and decay are governed by the principle of the minimization of the work, or the minimization of the difference between the rest mass energies and the kinetic energies of the particles involved.

$$\delta(m - m_o) = 0$$

We hypothesize that when generating high energy leptons, at some point it becomes more expedient for 'Nature' to generate a muon rather than an electron with a velocity $v/c \sim= 1$.

If true, where is this point? Is there a hard and fast rule governing when to choose the higher family particle, or is there some probabilistic indeterminacy involved?

In our model, the muon 'sheds' energy and spin in the form of its neutrino. The energy and spin shed ensures that the resulting virtual lepton propagator has spin = 0, and is *massless*.

This requirement places constraints on the energy and momentum of the initial and final states in muon (and tau) decays, and should help to explain the choice of the final state lepton from the three particle families in a particular decay, as well as some of the current mysteries surrounding lepton universality.

We can solve for this threshold, by assuming the muon rest mass is equivalent to the largest *allowable* relativistic mass of the electron.

$$m\_mu = m\_e/(1 - v^2/c^2)^{1/2} \qquad (22)$$

$$(v^2/c^2)_{THRESHOLD} = 1 - (m\_e/m\_mu)^2 \qquad (23)$$

A similar calculation will result in the velocity threshold between the mass of the muon and the mass of the tau.

We are not really clear on what "allowable" is: be it hard and fast or dependent on the particular dynamics of each interaction . . .

The leptonic table:

LEPTONS   ANTI-LEPTONS

electron	electron neutrino	PARITY ⇔	electron antineutrino	positron
⇐	CHARGE	MASS ↕	CHARGE	⇒
muon	muon neutrino	PARITY ⇔	muon antineutrino	anti-muon
⇐	CHARGE	MASS ↕	CHARGE	⇒
tau	tau neutrino	PARITY ⇔	tau antineutrino	anti-tau
⇔	mass isospin	charge isospin ↕	mass isospin	⇔

TABLE 1: The leptons and their interrelations; or the kleptogenesis of the leptoquarks.

Any lepton can be 'generated' from any other by the appropriate applications of the parity operator, the mass isospin operator, and our newly proposed 'charge isospin' operator.

Using various combinations of the step up and step down operators of SU(2) and SU(3), plus the parity operator, we can write any quantum mechanical interaction current in terms of the 'fundamental' neutrino neutral current.

The Schrodinger equation:

Our theory is a gauge theory, We will call it the universal gauge theory.   :)

Our 'gauge invariant' solution, as last explored in "On Matter, Mass, and Motion", now looks like this;

$$\psi = \exp(im_0c^2/\hbar\, t)\, \exp(-imc^2/\hbar\, t)\, \exp(i\mathbf{p}\cdot\mathbf{x}/\hbar) \quad (24)$$

$$\psi = \exp(\,i(\mathbf{p}\cdot\mathbf{x} - (m-m_0)c^2 t)/\hbar\,) \quad (25)$$

The relativistic Schrodinger equation is obtained by replacing the operator $p^2/2m_0$ with the operator p^2/m.

Kinetic energy:

$$pc = ch/\lambda \quad (26)$$

$$ch/\lambda = ch\nu/v \quad (27)$$

$$ch\nu/v = c\hbar\omega/v = Ec/v \quad (28)$$

$$E = mc^2 \quad (29)$$

A little algebra yields

$$E = pv = mv^2 \quad (30)$$

math is fun!

Reverse pilot wave theory:

In our theory, particles follow a well defined path determined by the momentum vector. The average momentum of a particle is obtained by squaring the wave function in the usual way.

To obtain the instantaneous or "interaction" momentum at any given point, we must use the real part of the particle wave function.

$$\psi = \sin(px - Et)/\hbar \tag{31}$$

$$\partial\psi/\partial x = p/\hbar \cos(px - Et)/\hbar \tag{32}$$

We define the velocity operator

$$\mathbf{V} \equiv -i\hbar/m \; \partial/\partial x \tag{33}$$

For a free particle, $x_0 = 0$, $v_0 = v$.

$$x = v_0 t \sin(px - Et)/\hbar \tag{34}$$

$$p = mv_0 \cos(px - Et)/\hbar \tag{35}$$

The reverse pilot wave idea is presented schematically in Figure 1.

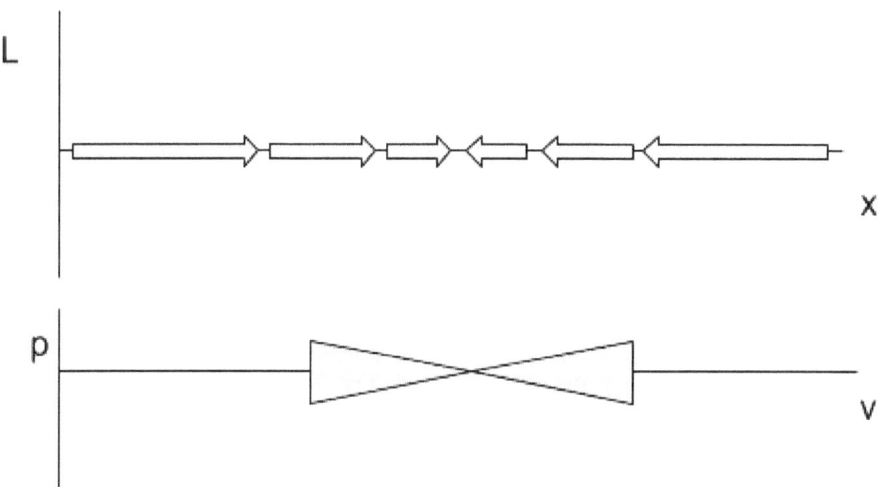

Figure 1: Reverse pilot wave theory.

The Dirac equation:

The universal Dirac equation is

$$H \psi = \alpha \cdot p \, \psi \qquad (36)$$

$$i\hbar \, \partial \psi / \partial t = -i\hbar \, \alpha \cdot \nabla \psi \qquad (37)$$

and is satisfied by, or solved with, the universal wave function, equation (25).

The hydrogen atom:

To date, we've been insisting that the electron orbits of the hydrogen atom are, or will be found to be, planar and circular. The straw man model. The actual orbits will be elliptical as determined by current quantum mechanics, however, the orbits are definitely planar.

Hydrogen orbits *must be* planar in order to conserve the constant first three integrals of the motion.

In our opinion, people determined an electron cloud, because they assumed, *beforehand*, the solutions would be spherically symmetric. One would not make such an assumption concerning the planetary orbits, although, it could be an interesting exercise.

In conclusion, the electron cloud is due to an honest mistake in judgement.

In the universal model, the hydrogen atom Schrodinger equation is the quantum analog of equation (7) which is our classical equation describing celestial orbits.

We also suggest the use of generalized coordinates, such as (θ, ω), and the corresponding uncertainty relationship

$$\Delta L_z \, \Delta \theta \geq \hbar / 2 \qquad (38)$$

Hydrogen atoms do not radiate in stationary states, because acceleration is only a necessary condition for radiation. Accelerated particles will only radiate, if there is also a change in *angular momentum*.

As an electron makes an orbital transition, or *accelerates*, it releases one unit of angular momentum, the photon, with the appropriate energy, etc.

We can now understand why classical electromagnetic waves are able to leave a source at the speed of light, while the sources themselves oscillate at much lower frequencies and speeds.

As one can see, our model explains *everything*, but spinning buckets !   :)

We don't do buckets.

Finally, the correspondence principle is easily explicable. The energy difference between photons emitted by electrons occupying very large orbits, becomes smaller and smaller, until the spectrum looks continuous and resembles classical electromagnetic waves.

However, emission is always discontnuous, and electromagnetic waves aren't real.

The proton:

In our model, the proton is a bound state of two positrons and an electron.

The strong coupling constant is (6)

$$\alpha_S = (G/4\pi\varepsilon)^{1/2} (2m_e e/\hbar c) (1 + \tfrac{1}{2} v^2/c^2 + \tfrac{3}{8} v^4/c^4 + \dots) \qquad (39)$$

We imagine the orbits and description of the particles of the proton to be the relativistic generalization of the current model for the molecule H_2.

The proton is essentially antimatter !

On math and physics:

All phenomena can be reduced to the interaction of mass currents.

Indeed, in our model, the fundamental particle of matter, the lepton, is a closed mass current spinning either to the left or the right. Particles spinning in the same direction repel, and those spinning in opposite directions, attract.

A spinning, three dimensional lepton, has three planar projections of spin angular momentum, **s • n,** resulting in an angular momentum per unit area, $\hbar/2$, along each of the three geometric axes of space.

Electric charge is essentially a second mass 'scale' that allows for the long distance interaction of the magnetic moments of two particles.

The point electric charge, e, is then the far field approximation of the electromagnetic charge, or dipole moment; $e \hbar/2 c$.
The same holds for the mass.

In our model, there is no renormalization, no infitities, no singularities, and no divergences; although we are not really sure what will happen with ultrasoft photon emission and the 'infrared problem'.

We replace Q^2, with v^2/c^2, or $\Delta V^2 = (v_1 - v_2)^2$.

The magnetic moment; a human-made, mathematical, contrivance, turns out to be the basis of all reality.

Imagine that.

Reformation:

Are we not all tired of scientific revolutions?

Let us consider Reformation, instead.

Since "Copenhagen", the field of physics has been overwhelmed by dull-witted, fantasy prone, magical thinkers.

Depending on one's taste, the results have been either comic or tragic.

Or, of no consequence at all; because, who really cares?   :)

Conclusion:

Dear friends,

This is probably the last book we will write about physics for a while;
of course, I think we've said this several times before!

I had a lot of fun, and I hope you did too.

Please take care of, care for, and care with, the universal model.

Enjoy!

Greg F

snake oil:

no embarrassment!
no shame

apologize not
or deign to explain

a physicist's credo
the psychics' creed

as priests
they steel
to the nave
turn heel !

to hide behind
sanctimony

duck blinds
of ceremony

still scrivening
their Impenetrable screeds

until they bleed

all coffers dry

questors heroic *!*

    questors heroic *!*
    tilt no more !

    toward
    the towering

    but towers !

    of ivory
    and babble

    bas relief
    in bronze

    canons and saints
    a chalice unknown

    pontiffs proclaiming
    ecclesiastically

    veneration

    of some dead
    philosopher's bone

I am:

    I am
    biology

    body and brain
    adrenaline
    bile

    sinew and sense
    pain trial

    homo sapien
    evolution's child

    reflex and reflection
    sight denial

    desire and doubt
    water loam

    a writhing

    wriggling
    flesh

    withering

    on recalcitrant bones

                          I am biology

                                          exploring itself

Resources:

Quantum Field Theory
Claude Itzykson, Jean-Bernard Zuber

Atomic and Quantum Physics
H. Haken, H.C. Wolf

Modern Elementary Particle Physics
Gordon Kane

Classical Dynamics of Particles and Systems
Jerry B. Marion

Foundations of Electromagnetic Theory
John R. Reitz, Frederick J. Milford, Robert W. Christy

Quantum Physics
Rolf G. Winter

Gauge Theories in Particle Physics
I. J. R. Aitchison and A. J. G. Hey

Quarks and Leptons: An Introductory Course in Modern Particle Physics
Francis Halzen, Alan D. Martin

Quantum Field Theory
F. Mandl, G. Shaw

Theoretical Mechanics of Particles and Continua
Alexander L. Fetter, John Dirk Walecka

Elementary Modern Physics   (Best Book Ever!)
Richard T. Weidner, Robert L. Sells

Books by Greg Feild: The SInister Universe Series

the pentateuch

1. "A quantum mechanical theory of gravitational interactions"
   CreateSpace Independent Publishing, 8/29/2016

2. "Observations on the quantum mechanical nature of gravity"
   CreateSpace Independent Publishing, 10/8/2016

3. "On gravitation and electric charge"
   CreateSpace Independent Publishing, 10/29/2016

4. "On spin, mass, and charge"
   CreateSpace Independent Publishing, 11/29/2016

5. "On angular momentum, acceleration, and absolute motion"
   CreateSpace Independent Publishing, 1/1/2017

the exegeses

6. "The Sinister Universe"
   CreateSpace Independent Publishing, 3/1/2017

7. "On Parity and Isospin"
   CreateSpace Independent Publishing, 4/11/2017

8. "Reflections on the Sinister Universe"
   CreateSpace Independent Publishing, 5/12/2017

the hermeneutics

9. "On Current Physics"
   CreateSpace Independent Publishing, 6/11/2017

10. "A Critical Examination of Classical and Quantum Mechanical Waves"
    CreateSpace Independent Publishing, 6/18/2017

the gospels   :)

11. "On wave particle duality and the quantum of action"
    CreateSpace Independent Publishing, 7/6/2017

12. "On matter, mass, and motion"
    CreateSpace Independent Publishing, 9/14/2017

13. "On action and reaction"
    CreateSpace Independent Publishing, 9/24/2017

14. "A quantum mechanical theory of everything"
    CreateSpace Independent Publishing, 11/5/2017

the compilations

"The Universal Model of Our Sinister Universe: The First Ten Books"
CreateSpace Independent Publishing, 7/2/2017

"The Canons of the Sinister Universe:
The Last Four Books on the Universal Model of Our World"
CreateSpace Independent Publishing, 11/5/2017

the expositions

15. "On Interaction"
    CreateSpace Independent Publishing, 4/21/2018

16. "On Rotation"
    CreateSpace Independent Publishing 8/19/2018

Learning physics from your dog:

Dogs know all about gravity. Dropped food falls to the ground.

Dogs also understand acoustics. The sound of plopping food will bring a dog running from rooms away.

If only I could teach my dog trigonometry.

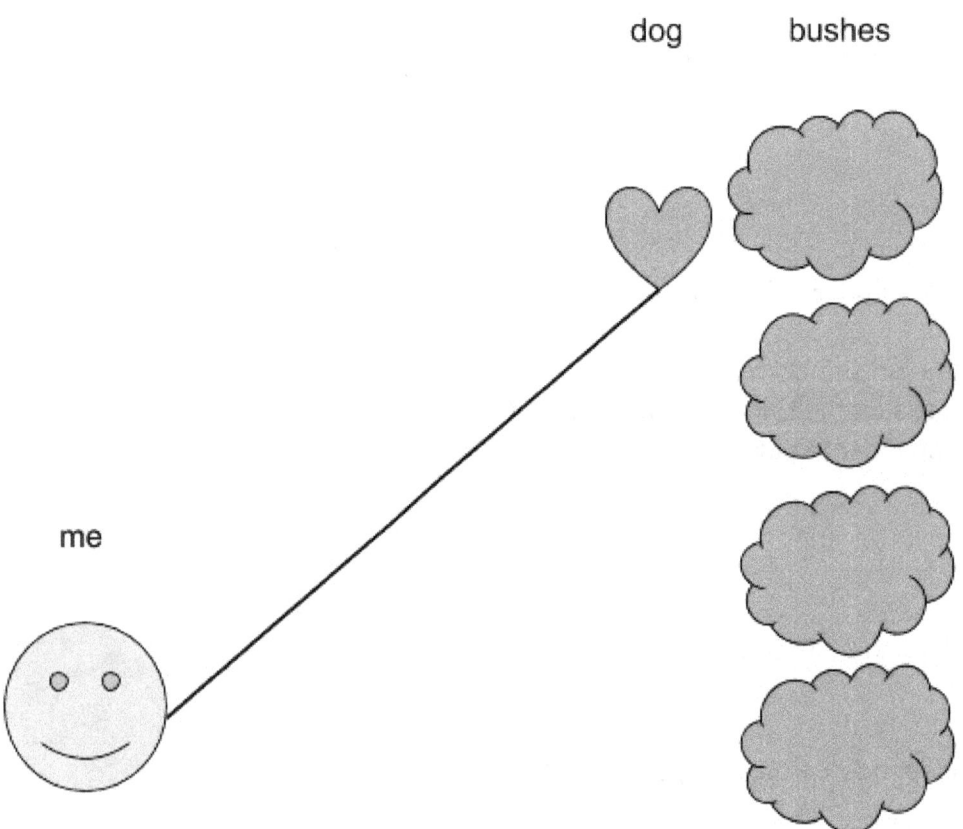

math is hard